Molecular Markers
in
Plant Genetics
and
Biotechnology

Molecular Markers in
Plant Genetics
and
Biotechnology

Editor
Dominique de Vienne
Institute National de la Recherche Agronomique
Versailles, France

CRC Press
Taylor & Francis Group
Boca Raton London New York

CRC Press is an imprint of the
Taylor & Francis Group, an **informa** business
A SCIENCE PUBLISHERS BOOK

CRC Press
Taylor & Francis Group
6000 Broken Sound Parkway NW, Suite 300
Boca Raton, FL 33487-2742

First issued in hardback 2019

ISBN 13: 978-1-57808-239-1 (hbk)

Ourage publié avec le concours du Ministère français chargé de la culture-
Centre national du livre.
This work has been published with the help of the French Ministère de la
Culture - Centre national du livre.

Library of Congress Cataloging-in-Publication Data

Marquerurs moléculaires en génétique et biotechnologies végtales.
English.
  Molecular markers in plant genetics and biotechnology/editor
Dominique de Vienne.
     p. cm
    Includes bibliographical references (p.).
    ISBN 1-57808-239-0
    1. Plant genome mapping. 2. Genetic polymorphisms. I. Vienne, D.
de (Dominique) II Title.
    QK981.45 M37 2003
    52.8'62—dc21
                                                      2002036561

Translation of: les marqueurs moléculaires en génétique et biotechnologies
              végétales, INRA, Paris, 1998. Text updated by author for the
              English edition in 2002

French edition: ©INRA, Paris, 1998
                ISBN 2-7380-07766-7
                ISSN 1144-7605

Visit the Taylor & Francis Web site at
http://www.taylorandfrancis.com

and the CRC Press Web site at
http://www.crcpress.com

Foreword

Does Science progress essentially by the acquisition of new tools or by the emergence of new concepts? This debate between *Homo faber* and *Homo sapiens* can be fed with several examples illustrating the relevance of these two points of view and, above all, the dialectic at work between these two sources of rejuvenation of approaches. Those who favour universal attraction, the theory of evolution or general relativity, are opposed to those who emphasize the decisive role of optic instruments (telescopes and microscopes of all types) or analytical tools (mass spectrometry, electrophoresis, magnetic resonance, etc.).

In this debate, molecular markers provide a fresh illustration of the key role that new tools can play in reviving scientific fields that are rich, even superabundant, in theoretical concepts but deficient in the capacity for refined observation of their object. Studies in quantitative genetics, as well as in population genetics, have progressively constructed a theoretical apparatus that often allows the explanation of a given fact by several hypotheses that are not mutually exclusive and that are often difficult to organize into a hierarchy. The 'genotype', which these disciplines constantly refer to, can be compared to a puzzle of which one knows only a few pieces and a very fuzzy overall image. It is clear from this analogy that debates on what the image represents could get lively!

Now, within a few years, geneticists have been possessed of a significant number of new pieces of the puzzle, and this new deal allows them to discriminate between hypotheses, or perhaps discard them all to construct new ones, illustrating thus the dialectic mentioned above between tool and concept.

The technological revolution has been extremely rapid. When DNA fingerprinting in humans was developed in the mid-1980s, many thought that its cost and complexity would limit it to criminology or specific applications. Twenty years later, it is difficult to study the genetics of an animal or plant species, to describe its biodiversity or for genetic improvement, without using molecular markers.

This volume, edited by Dominique de Vienne, very clearly presents these different approaches and their fields of application. The challenge was difficult: the techniques are rapidly developing and are often similar, and the definition of their 'field of relevance' requires a fine understanding of the problems that must be tackled. The authors seem to have succeeded thoroughly. Beyond its technical value, I am sure that this work will impart to the reader the present feverish enthusiasm of geneticists, who are coming closer and closer to revealing the traits of an object that they have long manipulated without really having known: the genome of plant and animal species.

Bernard Chevassus-au-Louis

Preface

The intensive use of molecular markers, even though it does not coincide with the emergence of a new scientific field, completely overturned several scopes in biology in the 1990s. One example is positional cloning, by which a gene with an unknown product can be isolated, just from its position in the genome. But this aspect, which medical genetics has widely advertised, must not obscure other equally fundamental applications. In the field of evolution, comparative mapping allows us to locate the genomic structural alterations that have occurred during the diversification of a genus or family. In formal genetics, the factors that influence the recombination rate can be analysed systematically. In population genetics, diversity and gene flow can be measured. The field most rejuvenated by molecular markers is quantitative genetics. Considered at first a "science for engineers", because it was used more by plant and animal breeders than by researchers, it is today successful in areas ranging from the analysis of genetic bases of morphological variation to the development of strategies for characterizing "quantitative" genes (the well-known QTL, for quantitative trait loci), and marker-assisted selection has become a powerful tool for plant and animal improvement. The recent complete genome sequencing of two model species, *Arabidopsis* (a dicot) and rice (a monocot) will greatly increase the potential of marker approaches in all higher plants, due to gene homology and synteny conservation in the plant kingdom.

To the extent that the set of methods known as 'molecular markers' apply to widely varying fields in genetics and plant biotechnology, it has not been possible to present the fundamentals of all the fields concerned. In each case we have wanted to show what specific inputs markers offer, the new questions they enable us to resolve, and possibly the problems they raise. The reader will need to refer, wherever necessary, to the fundamental works in genetics (formal, quantitative, or population genetics). Sound notions of molecular biology and statistics will also be useful. One exception is nevertheless made for the construction of genetic maps, the theory of which has been detailed for two reasons. First, the maps are the basis of most applications of markers; inadequate knowledge of their nature and way of construction would limit the reader's understanding of the rest of the book. Second, even though their principles have been well understood for 70 years, their teaching at the university level has been limited for decades for lack of new applications. Finally, the use of informatics offers new perspectives that merit attention.

The book is thus organized in the following manner. The first chapter details the different techniques of molecular markers, emphasizing genetic aspects, because these determine the type of use one can put it to. The construction of genetic linkage maps is the subject of the second chapter, where the advantages and disadvantages of the most common mapping populations are specified. The particular case of mapping of major genes, especially for the purpose of positional cloning, is addressed in the third chapter. Detection and applications of QTL controlling the expression of quantitative traits are presented in the fourth chapter, which also tackles the complex question of their identification. The fifth chapter underscores the major contribution of molecular markers in analysis of the structure and evolution of natural populations. Finally, the advantages of markers in selection, for studies of diversity and in the context of marker-assisted selection, are discussed in the last chapter. Throughout the book we have tried not only to expose in detail the principle of the numerous possibilities offered by markers, but also to illustrate them systematically with relevant bibliographical results. We have sought to give some unity to the work while respecting the spirit of each chapter, which is determined by the personality of the authors as well as the specificity of the field explored.

We hope that the work will therefore answer the needs of students as well as teachers, and that it will be a valuable tool for all researchers using markers, no matter what their objectives are.

Dominique de Vienne

Contents

Introduction

DEFINITIONS

In this book, the term *marker* is used in the sense of *genetic marker*. It is always synonymous with *marker locus*. A marker locus is a polymorphic locus that indicates:

- The genotype of the individual that carries it: it is for this purpose that markers are used in population genetics.
- The genotype of one or several loci linked to the marker; the applications here range from positional cloning to marker-assisted selection.

According to established terminology, the most commonly used genetic markers are *morphological* markers, *molecular* markers (at DNA level), and *biochemical* markers (isozymes, proteins). The last term is unfortunately ambiguous, since in other contexts it designates molecules, the presence of which indicates a differentiation stage or a physiological state. As we will presume the major principles of isozyme analysis are known to the reader, we will present essentially molecular markers. However, the markers from two-dimensional protein electrophoresis will be mentioned.

WHAT IS A "GOOD" GENETIC MARKER?

An "ideal" genetic marker is:

- Polymorphic: the geneticist's "raw material" is variability.
- Multiallelic.
- Codominant: a heterozygous hybrid simultaneously presents the traits of the homozygous parents; in a progeny, the heterozygotes can be distinguished from each of the homozygotes.
- Non-epistatic: its genotype can be inferred from its phenotype, whatever the genotype at other loci may be. Codominance and non-epistasis can be defined respectively as the absence of intra- and inter-locus interactions.

- "Neutral": the allelic substitutions at the marker locus do not have phenotypic or selective effects. Almost all molecular polymorphisms are neutral.
- Insensitive to environment: the genotype can be inferred from the phenotype, no matter what the environment is.

Morphological markers do not adequately meet these criteria. They are insufficiently polymorphic and are generally dominant. Besides, they often interfere with other traits and can also be influenced by the environment. Even though they are numerous in certain species (several hundreds in rice or maize), few of them can be jointly polymorphic in a given progeny.

Most biochemical or molecular markers, on the other hand, have all these qualities. The major limitation of isozymes is the small number of loci that can be detected. For instance, it is very difficult to have more than 30 to 40 isozyme markers in segregation in a rice or maize population. Also, all the enzymes are not present or active in all the organs. The polymorphic proteins detected by two-dimensional protein electrophoresis can be more numerous, but also depend upon the organ under consideration. In contrast, markers at the DNA level are nearly infinite in number and are independent of the stage or organ analysed, since DNA is the same in all the tissues. Moreover, they have the advantage that they can be more directly used for further applications in molecular biology.

Principal Sources of Molecular Markers

D. de Vienne, S. Santoni and M. Falque

CRITERIA OF CLASSIFICATION

Many molecular marker techniques are available today, and new ones are constantly published (see glossary at the end of this chapter). The acronyms are numerous, sometimes redundant, and often incomprehensible, which may confuse the uninitiated scientist. The aim of this chapter is not only to describe the principle of these techniques, but also to put them in some order. Although several criteria of classification are possible, we have chosen to emphasize genetic and molecular criteria, rather than technical or historical ones.

Genetically, we can distinguish between the techniques providing codominant and individually detected markers that satisfy the criteria of "good" genetic markers, as defined in the introduction, and those that provide patterns of multiple dominant markers (Table 1). This distinction is certainly simplistic; for example, one can find locus-specific and dominant markers. Nevertheless, the classification does correspond to two major uses of markers.

Molecularly, we can classify polymorphism into two categories: (1) *sequence* polymorphism (including nucleotide substitutions and insertions-deletions), and (2) *number of tandemly repeated sequences* in the repeated regions. There are no techniques developed specifically for insertion-deletion polymorphism, which is in any case detected by techniques that detect nucleotide substitutions. We will therefore consider on the one hand sequence variations, mentioning where needed the insertion-deletion variations, and on the other hand the variations of number of tandem repeats. Combining the genetic and molecular criteria, we get the four groups of markers presented in Table 1.

Table 1. Classification of molecular marker techniques
(for explanation of acronyms, see glossary)

Genetic criterion	Molecular criterion			
	Sequence (including insertion-deletion)[a]		Number of repeat units in repeated DNA	
	Searched difference	Technique	Size of the repeat unit	Technique
Codominant markers, detected individually	Site of restriction enzyme (RE)	RFLP and CAPS	1 to 4 nucleotides (microsatellite, or SSRs)	Acrylamide or agarose electrophoresis
	Conformation	SSCP		
	Stability	D/TGGE		
	SNP	– Primer extension – DHPLC – Pyrosequen- cing – DASH, etc.		
Multiple dominant markers (genetic fingerprints)	Hybridization site of arbitrary primer	MAAP: – RAPD – AP-PCR – DAF	1 to 4 nucleotides (microsatellite, or SSRs)	ISSR (SSR primer + some arbitrary bases)
	RE sites and arbitrary primer	AFLP	5 to > 100 nucleotides (minisatellites)	Southern with minisatellite probe

[a]Apart from the particular case of repeated DNA (right-hand column), there is no technique to detect *specifically* insertion-deletion polymorphism, which is a particular case of sequence polymorphism. When a polymorphism is observed, we cannot thus know what its origin is without supplementary experiments. From the genetic point of view, this ambiguity does not matter, the goal being to have polymorphic loci, whatever may be the origin of the polymorphism.

CODOMINANT MARKERS DETECTED INDIVIDUALLY

Sequence polymorphism

Sequence polymorphism can of course be visualized by sequencing homologous fragments. Despite progress in automation, this technique remains cumbersome, especially when the acceptable rate of error is very low. The sequencing of alleles is presently not a routine method when a large number of loci are being considered. We therefore use indirect methods, not exhaustive, but much faster and cheaper, based on the detection of differences in:

– Restriction sites
– Conformation
– Stability
– Sites of hybridization of oligonucleotides
– Length of fragments

DIFFERENCES AT RESTRICTION ENZYME SITES: THE RFLP TECHNIQUE

Restriction enzymes

Restriction enzymes, also known as restriction endonucleases, are enzymes that cut (or "digest") DNA at specific sites, called *restriction sites*, comprising most often an even number of bases (4, 6, or 8, sometimes more), generally arranged in a palindrome (Table 2). Thus, if the bases A, T, G, and C of the DNA were equally frequent, an enzyme having a recognition site of 6 bases would cut the DNA on average every 4096th base (4^6). A genome of 10^9 bases will thus produce around 250,000 fragments of variable lengths. The specificity is such that the replacement of a single base in a site is enough to prevent the enzyme used from cutting the DNA. It is this specificity that is exploited for the detection of a polymorphism: a presence or absence of the restriction site leads to a *fragment-length polymorphism* (Fig. 1). This

Table 2. Restriction enzymes currently used for RFLP

Emzyme	Site de restriction
Apa I	5'... GGG CC|C ...3' 3'... C⌐CC GGG ...5'
Cla I	5'... AT⌊CG AT ...5' 3'... TA GC⌉TA ...5'
Bam HI	5'... G|GATC C ...3' 3'... C CTAG|G ...5'
Dra I[1]	5'... TTT |AAA ...3' 3'... AAA| TTT ...5'
Eco RI	5'... G⌊AATC ...3' 3'... CTTAA|G ...5'
Eco RV[1]	5'... GAT |ATC ...3' 3'... CTA |T AG ...5'
Hind III	5'... A⌊AGCT T ...3' 3'... T TCGA|A ...5'
Pst I[2]	5'... CTGCA|G ...3' 3'... G⌐ACGTC ...5'
Pvu II[1]	5'... CAG|CTG ...3' 3'... GTC|GAC ...5'
Sma I[1]	5'... CCC|GGG ...3' 3'... GGG| CCC ...5'
Xba 1	5'... T⌊CTAGA ...3' 3'... A GATC|T ...5'
Xho I	5'... G|AGCTC ...3' 3'... CTC GA|G ...5'

(1) Unlike other enzymes, which generate "cohesive ends" by cutting, *Dra* I, *Eco* RV, *Pvu* II, and *Sma* I generate "blunt ends".

(2) When 5' cytosines are methylated, *Pst* I does not cut this restriction site. This property is used for the construction of genomic libraries enriched in sequences that are seldom or not repeated.

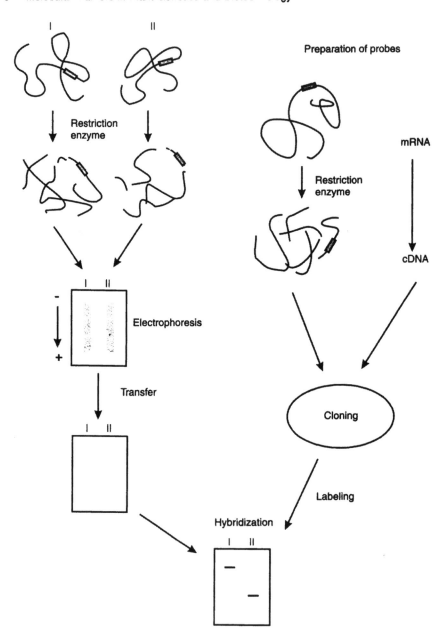

Figure 1. Principle of detection of restriction fragment length polymorphism (RFLP). I and II represent two individuals. A filament schematizes the DNA molecule. The gray rectangle on the DNA represents a particular region recognized by a probe. After digestion, the fragment containing this region is longer in individual I than in individual II. This difference in length will be detected by hybridization of the labelled probe on the DNA after migration (see text for more details). This probe thus reveals a biallelic locus (Botstein et al., 1980).

phenomenon is not rare. Since polymorphism is a basic property of living beings, digestion of the DNA of any two individuals in a given species produces a great number of differences of fragment length. Although it is easy to obtain polymorphic fragments, it is difficult to see them. In fact, electrophoresis of the products of digestion, in agarose or acrylamide gels (see box 1.1), followed by a detection with ethidium bromide or silver nitrate, does not allow us to distinguish between individual fragments,

---

Box 1.1: Selection of a migration medium for detection of nucleic acid polymorphism: agarose or acrylamide?

The choice between acrylamide and agarose as a medium for migration of nucleic acids depends on several factors.

Practically, agarose gels are easier to make and use than acrylamide gels. In the first case, we simply melt the agarose in a suitable buffer solution and run the gels in relatively simple migration tanks. Acrylamide gels, which run between two glass plates, require a more sophisticated and expensive migration apparatus.

The choice of a migration medium will depend above all on the level of resolution required. At usual concentrations (around 6%), acrylamide polymerizes in a network, the high separation power of which allows us to distinguish between fragments that differ only by a few base pairs, sometimes just one or two. This resolution is exploited particularly for detection of microsatellites, otherwise known as simple sequence repeats or SSR, as well as for certain genetic fingerprinting techniques (DAF, AP-PCR, AFLP™, sometimes RAPD) and STS (SSCP and D/TGGE) (see glossary). In gels of usual size, the readable zone does not exceed 500 bp. Separation on agarose gel (concentration around 1%) enables us to extend the range of fragment sizes that can be observed to the maximum, from 300 to 15,000 bp, but with a power of separation of only around 50 bp or more. Agarose gel is well adapted to the RFLP technique and makes it possible to transfer significant quantities of DNA onto the nitrocellulose or nylon membranes for probe hybridization.

Automatic sequencers now make electrophoresis on acrylamide medium even more effective. In such devices, fluorescently labelled DNA samples are detected within the gel during migration, and fluorescence intensity is recorded along time with a laser at a given migration distance, so that the size of the molecules is estimated by the time required until it reaches the detector. Compared to classical polyacrylamide gel electrophoresis devices, automatic sequencers avoid the time-consuming steps of gel staining and reading after migration and make automatic data acquisition easier. In addition, such systems increasingly tend to use capillary electrophoresis instead of slab gels. Samples are then run in individual capillaries filled with a polyacrylamide-like polymer, instead of being loaded side by side in the same flat gel. The electrophoresis performances of capillary systems are not yet quite as high as those of slab gels, but filling the capillaries with polymer and loading samples can be completely automated, which gives a great advantage to capillary sequencers for high-throughput analyses.

because they are much too numerous, causing a smear of coloration along the migration lane. On the other hand, even when visualization is possible, the problem of locating homologous fragments remains. In genetic terms, how do we distinguish allelic fragments from non-allelic fragments? The use of molecular probes can resolve these problems.

Molecular probes

Molecular probes are small fragments (0.5 kb to 3 kb) of DNA or RNA that can be used to locate a particular DNA within a complex mix. This is possible because these complementary DNA strands have the capability of *hybridizing*. Even on a very long genome, a given probe will hybridize only its complementary sequence, and nowhere else. Provided the probe is tagged radioactively or chemically, the desired fragment can be identified.

Two sources of probes are commonly used, *genomic DNA* probes (gDNA) and *complementary DNA* probes (cDNA).

Genomic DNA probes are obtained by digestion of total DNA of the nuclear genome of the species studied with the help of a restriction enzyme. To obtain fragments of length compatible with techniques of cloning or *in vitro* amplification, an enzyme with 6 bases in its recognition site is generally chosen. But the genome of higher plant species contains significant quantities of repeated DNA. Moreover, this DNA is often not expressed, so that the technique produces many redundant fragments; most of them do not correspond with genes. To overcome this disadvantage, one can use restriction enzymes sensitive to methylation (Table 2). The idea is to exploit the fact that the regions expressed are much less methylated than the repeated, often silent, regions. An enzyme such as *Pst* I will therefore generate very large fragments in the methylated repeated regions and smaller fragments in the regions with genes that are seldom or never repeated. The libraries of genomic DNA probes obtained with *Pst* I are thus usually "enriched" in single copy DNA sequences as compared to conventional genomic libraries (Table 3).

The cDNA probes necessarily correspond to expressed genes. Messenger RNAs (mRNAs) are extracted from a given organ, their complementary DNA is synthesized with the help of an enzyme, reverse transcriptase, and the cDNAs are cloned to be used as probes. Compared to gDNA probes, cDNA actually reveal a high proportion of unique loci, whatever the species considered (Table 3).

Table 3. Percentage of unique loci detected according to origin of probes

Species	gDNA – *Eco* RI	gDNA – *Pst* I	cDNA
Tomato	33	92	95
Rice	50	58	85
Lentil	38	41	88

In practice, *Pst* I probes and cDNA probes are the most commonly used. As we shall see later, these two categories of probes sometimes have different fields of application.

The steps of the technique

The simplified RFLP technique comprises the following steps (Fig. 1):

1. DNA of different genotypes to be analysed is extracted.
2. The DNA is digested by a restriction enzyme.
3. The fragments created are size-separated by agarose gel electro-phoresis. The DNA, being negatively charged, migrates from the cathode to the anode. The smallest fragments are the quickest. (Conventionally, the gels are represented with the origin of the migration at the top.)
4. The DNA is transferred in denatured form to a nylon membrane; the relative position of DNA fragments is preserved during the transfer.
5. The membrane is incubated in the presence of a solution containing a probe previously tagged, either radioactively or chemically (see box 1.2). The probe then hybridizes with the DNA fragment or fragments with which it is homologous.
6. The site or sites at which the probe is hybridized are revealed by placing the membrane in contact with a radioactivity-sensitive film, or by staining with a specific enzymatic reaction (Fig. 2).

Box 1.2: Probe preparation: radioactive and cold probes

In order to determine where a probe is fixed on a migration lane, the probe must be tagged, either radioactively or chemically. In both cases the basic principle most often used is the same, *random priming*. The probe, in the single strand form, is put in the presence of a DNA polymerase (Klenow enzyme), of free nucleotides, and of a combination of small oligonucleotide primers (6 bases) of arbitrary sequences. These will hybridize to the probe DNA each time they find a complementary sequence, thus creating points of initiation for DNA synthesis by the polymerase. Some nucleotides can be labelled radioactively or linked to an antigen. In the first case, their incorporation in the probe makes it radioactive, while in the second case one uses an antibody combined with an enzyme such as alkaline phosphatase to visualize the band with the help of a coloured reaction ("cold" probe). The cold probes are presently as sensitive as radioactive probes.

Cold probes are not always cheaper than radioactive probes. They can, however, be used in any laboratory, unlike radioactive probes, which demand well-equipped and authorized laboratories and which must not be handled in large quantities for safety reasons. Moreover, cold probes can be stored at −20°C for several months, whereas radioactivity decays over time. Nevertheless, it must be noted that the chemistry used in cold probe preparation is far from harmless and, therefore, rigorous precautions must also be followed.

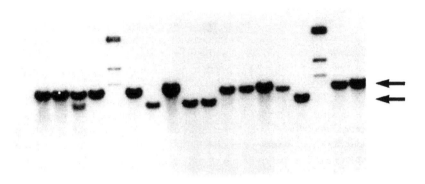

Figure 2. Example of result of an RFLP experiment in a population of recombinant inbred lines of maize. Lanes 1 to 16 correspond to different individuals of the progeny. M: molecular weight standard. All the individuals are homozygous, except individual 3. This residual heterozygosity arises from an insufficient number of self-fertilizing generations.

Steps 4 to 6 correspond to the technique of Southern (1975). Botstein et al. (1980) published the complete RFLP technique.

Polymorphism detected

When two genotypes, which we will suppose to be homozygous, are compared after digestion of their DNA by a given enzyme and hybridization with a given probe, three elementary situations may occur:

a) Near the region of the genome recognized by the probe, the restriction sites are the same, and there are no insertion-deletion differences. The patterns are identical (Fig. 3a).

b) There are differences of restriction sites, without differences in insertion-deletion. The patterns are different (Fig. 3b).

c) There are differences in insertion-deletion, without differences of restriction sites. The patterns are different (Fig. 3c).

Situations b and c are not mutually exclusive. It must be noted that in the presence of different patterns, it is not possible, without complementary experiments, to distinguish between situations b and c. However, the use of several restriction enzymes enables us to estimate the relative proportion of restriction sites and insertion-deletion, because an insertion-deletion in a region recognized by a probe will be revealed by any enzyme, whereas there will be no correlation between enzymes in case of restriction site polymorphism. With polymorphism data for a large number of probes and at least two enzymes, a statistical test will enable us to say whether there is a significant percentage of insertion-deletion in the species considered (see box 1.3). This type of test revealed, for example, that insertion-deletion polymorphism is significant in maize and negligible in tomato.

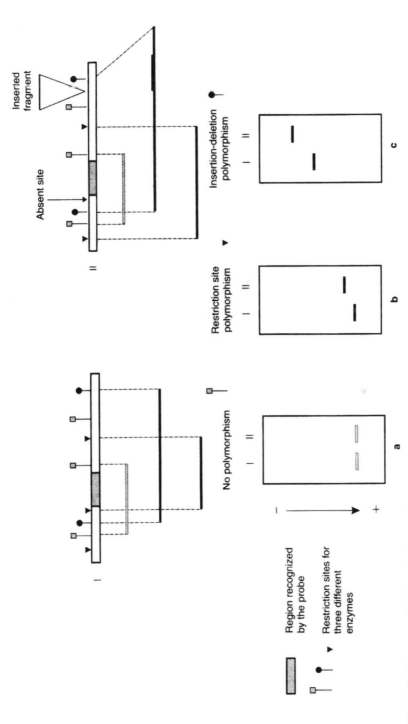

Figure 3. Elementary situations of polymorphism revealed by RFLP technique between two individuals, I and II. a: No polymorphism. b: Restriction site polymorphism. c: Insertion-deletion polymorphism. The restriction sites of three enzymes near a region recognized by a probe (in gray) are represented by different symbols.

Box 1.3: Test for presence of insertion-deletion polymorphism: the example of pea

In this experiment, 166 genomic probes were obtained by digestion of pea DNA with the enzyme *Pst* I. To discover whether these probes reveal polymorphism between the two varieties "Erygel" and "661", their DNA was digested with the two enzymes *Eco* RI and *Hind* III. The probes were classified into the following four categories according to the polymorphism detected: polymorphism with *Eco* RI and with *Hind* III, with *Eco* RI but not with *Hind* III, with *Hind* III but not with *Eco* RI, and with neither of the two enzymes. The numbers of each type of probe were indicated, as well as the subtotals (Σ) and the total (T). In parentheses are indicated the numbers expected in the case of independence between enzymes, that is, when the fact that a polymorphism is detected with one enzyme has no effect on the probability that it will be detected with the other enzyme.

	Hind III		
Eco RI	Monomorphic	Polymorphic	Σ
Monomorphic	84 (66.4)	21 (38.6)	105
Polymorphic	21 (38.6)	40 (22.4)	61
Σ	105	61	T = 166

The χ^2 test of independence (1 degree of freedom) gives the value 34.5, which is highly significant ($P \leq 0.001$). If this result, obtained with only two genotypes, is general, it indicates that there is a significant amount of insertion-deletion polymorphism in the pea genome, in addition to restriction site polymorphism (Dirlewanger et al., 1994). But this type of test does not enable us to quantify the relative proportions of these two sources of polymorphism.

The probe libraries are usually made of several thousands of clones. Before these probes are used to analyse polymorphism in a population, they must be screened on a sample of individuals, the DNA of which has been digested by some restriction enzymes. Several factors lead to the rejection of a large number of probes. The probes that do not give a signal, that produce considerable "background noise" or complex patterns, or that do not reveal polymorphism are rejected (Fig. 4).

This latter factor can be crucial for certain species. For example, in the cultivated varieties of tomato, *Lycopersicon esculentum*, only 5% to 10% of probes reveal polymorphism after digestion by three restriction enzymes, with only two alleles per locus on average. This explains why one must often use interspecific crosses to map the genome of this species. On the other hand, considering only the lines of the dent group of maize, which represent only a fraction of the variability available in this species, 95% of the probes reveal polymorphism with three restriction enzymes, the mean number of alleles per locus being higher than 6. Actually, isozyme markers had already revealed that difference between the two species.

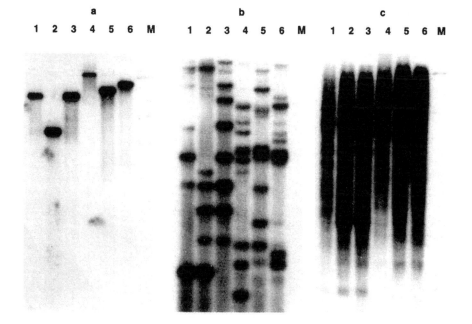

Figure 4, Screening probes for RFLP. The DNA of three maize lines has been digested by *Hind* III (lanes 1 to 3) and *Eco* RI (lanes 4 to 6). (M: molecular weight standard). The profiles obtained with three probes show different situations. a: The profile is simple, the probe reveals a single sequence in the genome. b: The profile is multi-band, the probe reveals several loci. c: The probe produces an intense background noise, because it corresponds to a highly repeated sequence in the genome.

Genetic aspects

From the genetic point of view, the type of molecular polymorphism is not important. A probe reveals a locus, which can be either polymorphic or monomorphic. When the number of enzymes becomes higher, the number of detectable alleles at this locus in a population also increases. In a heterozygous individual, the two allelic bands are visible. In other words, this technique enables us to distinguish clearly the heterozygote from both homozygotes. It is a source of *codominant* markers detected *individually*. However, in certain species that have undergone duplications of all or part of their genome during the course of evolution, a probe can reveal two or several unlinked loci. Provided the patterns are not too complex, genetic analysis will enable us to say which are the allelic bands, and it is usually not a problem for the construction of genetic maps. Besides, restriction sites can be found within the region recognized by the probe. There will, therefore, be competition between the different fragments for hybridization with the probe, and two or several bands will appear for a given locus, even in the homozygotes (Fig. 5). In the case of polymorphism, an allele can thus

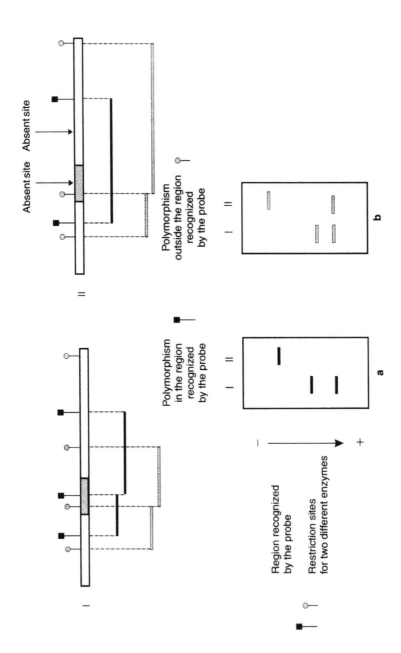

Figure 5. RFLP profiles observed when there is a restriction site within the region recognized by the probe. a: Enzyme with an internal polymorphic site and external monomorphic sites. b: Enzyme with an internal monomorphic site and an external polymorphic site. The fact that an enzyme cuts within the region recognized by the probe does not change anything in the genetic interpretation. Merely, one of the alleles (a) or both of them (b) appear as a two-band profile.

correspond to two or several bands. But there also, genetic analysis enables us to elucidate the exact allelic relationship.

Heterospecific probes

We use the term *heterospecific* probes for probes that do not belong to the species studied (the common term *heterologous* probe is improper, because it involves precisely homologous sequences). The percentage of usable heterospecific probes decreases as the genetic distance increases, and it is rare to be able to go beyond the level of the botanic family. With this consideration, cDNA probes are *a priori* more interesting because, arising from coding regions, they are more conserved through evolution, and therefore better preserve their capacity for hybridization. The gDNA probes obtained with *Pst* I, enriched in single sequences, are still quite effective, while gDNA probes non-enriched in single sequences are ineffective (Table 4). Practically speaking, heterospecific probes enable us to

Table 4. Efficiency (%) of heterospecific probes from oats and rice on Poaceae

	Oats	Rice	Wheat	Barley	Sugarcane
cDNA (oats)	100	83	65	71	60
gDNA (rice)	2	100	0	0	0

With a high rate of success, cDNA probes of oats hybridize with the DNA of species belonging to rather distant tribes (rice and sugarcane). They can therefore be used routinely, for example, for comparative mapping. In contrast, the gDNA of rice recognize the DNA of oats at a very low rate, and do not recognize those of other species at all. The differences between the two types of probes arise from different rates of evolution between coding and non-coding sequences.

map the genome of a species without having to repeat the considerable effort of isolating and sorting the probes. Moreover, they offer the possibility of comparing the maps of related species. Apart from its fundamental interest, comparative mapping is a valuable tool to facilitate the isolation of genes, as we will see in Chapter 2.

A PARTICULAR CASE OF RFLP: CAPS*

The polymerase chain reaction (PCR) technique, or *in vitro* enzymatic amplification of DNA, has made it much easier than before to amplify a given DNA fragment in any series of individuals (STS, sequence tagged site, see glossary). Various methods for analysing the polymorphism within a product of amplification are summarized in box 1.4 and are detailed later. One of them, called CAPS (or PCR-RFLP), consists in digesting the amplified fragment with one or several restriction enzymes having 4-base recognition sites and revealing the polymorphism at the restriction sites by agarose gel electrophoresis. The choice of enzymes with 4-base recognition

---

*All acronyms are explained in the glossary at the end of this chapter.

**Box 1.4: Detection of polymorphism within products of amplification
(STS, Sequence Tagged Sites)**

With the use of PCR (polymerase chain reaction) techniques, it is possible to amplify
a given DNA fragment *in vitro* by a cyclical enzymatic reaction, even if it is present as
a single copy within a large genome. To do this we must know the sequence of the
"targeted" fragment, or at least that of its ends on about 20 bases, in order to
synthesize the oligonucleotide primers of the reaction. The fragment length must not
exceed 2 or 3 kb, except under particular conditions of PCR. If we wish to search for
polymorphism *within* the fragment, we first use the pair of primers to amplify the
fragment in a series of individuals. Depending on the goal of the study, the type of
polymorphism sought, and the available resources, we can then implement one of the
following techniques (see text for more details):

- Direct electrophoresis, which reveals the insertion-deletion differences and the
  differences in the number of sequence repeats (essentially for microsatellites).
- Electrophoresis after digestion of the fragment by restriction enzymes (CAPS,
  cleaved amplified polymorphic sequence).
- SSCP (single strand conformational polymorphism).
- DGGE (denaturating gradient gel electrophoresis) or TGGE (temperature gradient
  gel electrophoresis).
- Several SNP genotyping techniques, which all aim at determining which
  nucleotide is present at a given site of the sequence of the PCR fragment. This
  approach requires knowing beforehand the SNP positions, that is, the nucleotide
  sites that display polymorphism.
- Sequencing, which reveals exhaustively all the substitutions and insertions-
  deletions, but which is hardly feasible as a routine technique.

Those markers arising from PCR amplification of a *specific* fragment are sometimes
designated as STS, for Sequence Tagged Sites. They are widely used, because they
combine the technical advantages of PCR and the genetic advantages of RFLP.
Unlike RFLP, they do not require previous digestion of the DNA, or transfer, or
preparation and tagging of probes (with some exceptions), or hybridization. Besides,
PCR can be automated, and the quantities of DNA required are much lower than for
RFLP. Genetically, they are a source of locus-specific codominant markers. The
limitation on their use is not so much the need for prior knowledge of sequences (more
and more cDNAs have been sequenced), as the cost of primers, the optimization of
PCR conditions appropriate to each STS, and the implementation of an efficient
polymorphism detection method.

---

sites (one 4-base site is expected every 256 (4^4) base on average), as imposed
by the length of products of PCR, which normally range between 0.5 and 2
kb, enables us to have a high probability of cutting. Like "classic" RFLP, this
technique provides codominant markers that are detected individually. In
plants, it has been used for classification of rice genotypes (Ghareyazie et
al., 1995), for gene mapping in barley (Tragoonrung et al., 1992), and for

screening STS linked to resistance genes in lettuce (Paran and Michelmore, 1993). Its applications in population genetics have been presented by Karl and Avise (1993).

DIFFERENCES OF CONFORMATION: SSCP

When a double-stranded DNA fragment is heat-denatured at 95°C, then quenched, the single-stranded molecules do not have time to reassociate in double-stranded molecules, but form a folded secondary structure stabilized by local double-stranded regions (Fig. 6). The differences in sequence can lead to differences of conformation that are detected by migration in non-denaturing conditions in acrylamide gels (Orita et al., 1989). It is estimated that nearly 100% of the differences in sequence can be detected in fragments smaller than 200 bp. This percentage decreases as the length of the fragment increases. In a homozygous individual, two bands can generally be observed, because the two strands of a given DNA molecule usually have slightly different secondary conformations. In some cases, several bands may appear because of the existence of several semi-stable conformations for the same strand. There are only a few examples of use of this technique in plants, for mapping or population genetics (Fukuoka et al., 1994; Bodenes et al., 1997). It is, however, potentially very useful, because it does not require digestion, like CAPS, or conditions of electrophoresis as precise as those of D/TGGE (see below). It is, therefore, useful for the rapid screening of variants of products of amplification, when precise knowledge of substitutions involved is not required.

DIFFERENCES IN STABILITY: D/TGGE

Differences in sequence can lead to *local* differences in stability of the double-stranded DNA molecule as a response to denaturing agents such as urea-formamide or temperature. This property is used in the techniques of DGGE and TGGE, respectively. After PCR amplification, the DNA fragment is loaded on an acrylamide gel. The conditions of gel preparation, or the conditions of migration, are such that the DNA molecule migrates along a denaturation gradient: either an increasing temperature or increasing concentration of urea and formamide. The DNA migrates first as a double strand, then meets conditions that denature the least stable complementary regions of the molecule, forming a branched structure that does not migrate further (Fig. 7). In most cases, and provided the fragment studied is not too long (< 300 bases), a difference of one base in the least stable regions leads to a difference of position of the band. Moreover, the heterozygous individuals can all be identified. This can be done by a denaturation step after the last PCR cycle, followed by a renaturation, which will form a heteroduplex. These have a very low annealing temperature and will thus give rise to bands that do not migrate very far.

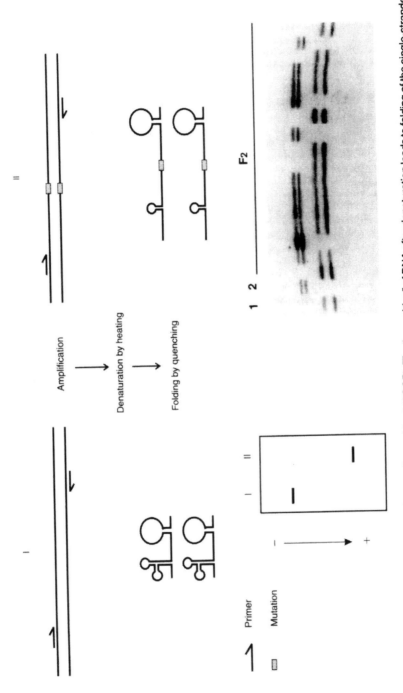

Figure 6. Single strand conformation profile (or polymorphism) (SSCP). The "quenching" of DNA after denaturation leads to folding of the single-stranded molecule. The photograph shows an example of polymorphism revealed in an F_2 population of rice. There are two segregating alleles, with two bands for each of them (after folding, the two strands of DNA do not migrate necessarily at the same rate). The heterozygotes therefore present 4 bands. Lanes 1 and 2 correspond to parents of the progeny (after Fukuoka et al., 1994).

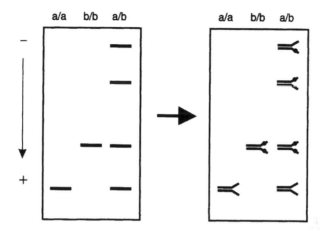

Figure 7. Denaturing gradient gel electrophoresis (D/TGGE). At left, bands observed for three genotypes *a/a*, *b/b*, and *a/b*. At right, interpretation: scheme showing partly denatured DNA, forming a branched structure. The point on the DNA indicates the position of a substitution between fragments *a* and *b*.

These bands are then easily detected, and even if a substitution cannot be detected by comparison of homozygotes, it can often be discovered with the help of heterozygotes.

Practically, the D/TGGE is a more delicate technique than SSCP. The gradient must be chosen according to the fragment to be studied. Appropriate software (Lerman and Silverstein, 1987) can determine the annealing temperature along the fragment and select the slope and limits of the gradient according to the least stable complementary regions. If the molecule does not present two distinct complementary regions, a *GC clamp*, that is, a region that contains only GC bounds, which are highly stable, can be added to the molecule. For this, one uses primers extended at the 5' end by about 30 C and G bases.

Because the migration of different variants of a given fragment can be modelled, this technique is well adapted for the quick screening of alleles in a population. However, that it requires precise knowledge of the sequence to be studied.

SNP GENOTYPING TECHNIQUES

The term SNP designates a nucleotide site of a given sequence for which substitution polymorphism has been observed at a significant frequency between different individuals. However, insertion-deletion polymorphisms (IDPs) are sometimes also incorrectly referred to as SNPs. The initial sequence information can be specially produced for SNP discovery by sequencing a given fragment in different genotypes, but can also be a by-product, for instance of large EST sequencing programmes performed with cDNA libraries from different genotypes. In the latter case, significant SNP discoveries can be carried out completely *in silico* from existing databases. The SNPs are then located in transcribed sequences, which is interesting for high-throughput mapping of genes, for instance in candidate gene approaches.

Like CAPS, SSCP, or D/TGGE, the SNP genotyping techniques listed below detect sequence polymorphisms within a specific PCR fragment (STS). In addition, they make it possible to visualize all the possible single-nucleotide changes within the fragment, whereas the previous techniques are applicable only if the SNP is part of a restriction site, or sufficiently changes the DNA strand's conformation or stability. Moreover, because of the growing need of high-throughput genotyping, the possibilities of automation and the quickness of the analyses are major criteria for the selection of SNP genotyping techniques. For this reason, electrophoresis is avoided as much as possible and replaced for instance by direct fluorescence or luminescence signal detection in the reaction tube.

The basic principles that are used to distinguish between the possible nucleotides at the SNP sites are based on:

- Allele-specific oligonucleotide hybridization (ASOH; related techniques: allele-specific PCR, 5' nuclease assay, DNA chips, bead-based techniques)
- Template-dependent DNA strand elongation by a DNA polymerase (related techniques: primer extension, pyrosequencing)
- Double-strand–dependent ligation (related technique: oligonucleotide ligation assay (OLA) coupled with DNA chips or bead-based techniques)
- Mismatch detection (related techniques: DASH, DHPLC)

Because no particular technique has yet broken through to take over all the other ones, we will describe the different approaches mentioned above, each of them being potentially convenient for some SNP genotyping project or other. This list is far from being exhaustive because a great deal of methodological research is presently being carried out on this subject, owing to its possibly significant economical return on investment.

Allele-specific PCR

The basic idea of allele-specific PCR is that PCR amplification is dramatically reduced if there is a mismatch at the last 3' base of one of the primers. It is then possible to design an allele-specific PCR test with the last 3' base of one of the primers on the SNP, and the other primer in a conserved sequence not too far from the first one. This simple presence/absence test can be performed directly on total genomic DNA, and the STS is amplified only for one of its alleles. Practically, things are not so simple for two reasons: (1) one mismatch at the 3' end of a primer is not always sufficient to prevent PCR amplification and (2) absence of a PCR product can also be caused by an unsuccessful experiment. To overcome the first point, a second mismatch can be added in the vicinity of the 3' end of the allele-specific primer to destabilize its priming end. This mismatch alone will still allow PCR amplification, but there will be almost no amplification if the first 3' base of the primer forms a second mismatch. To overcome the second point, it is possible to design one allele-specific primer for each allele, and to perform one PCR reaction for each allele. In case of unsuccessful experiment, none of the primers will amplify, so that a missing data cannot be considered as an allele. To avoid the multiplication of PCR reactions, single-tube PCR can be performed with four different primers designed so that each allele will give a different size of band. Ye et al. (2001) presented an example of this approach, as well as software tools to optimize primer design. Such analyses are simple to perform and require no particular expensive equipment, but relatively few SNPs can be genotyped successfully by this means without optimization, so that the overall throughput remains low and this technique is more suitable for projects involving a small number of SNPs.

5' nuclease assay (TaqMan™ assay)

This technique uses the real-time quantitative PCR method based on the 5' exonuclease activity of the Taq polymerase (Heid et al., 1996) to simultaneously perform PCR amplification and allele-specific oligonu-cleotide hybridization. Thus, the analysis is quick and can be easily automated. PCR amplification is performed in the presence of an oligonucleotide probe (TaqMan™ probe) designed to hybridize to the PCR product in the region containing the SNP. TaqMan™ probes are coupled at their 5' end to a reporter dye and at their 3' end to another dye that acts as a quencher. This means that as long as the two dyes are physically close to each other, the complex will not fluoresce, but if they are separated, the reporter dye will become fluorescent. During the elongation steps of the PCR reaction, the TaqMan™ probe hybridizes to the template strand, and as the Taq polymerase reaches the probe, its 5' exonuclease activity cuts the probe and thus releases a free fluorescent reporter dye separated from its quencher. Fluorescence intensity is monitored during PCR in a special thermal cycler. The TaqMan™ probes are designed with the SNP position close to their 5' extremity, so that if the probe presents a base at the SNP different from that of the template, the 5' end of the probe will not hybridize and the Taq polymerase will not be able to cut it and release the dye. For more reliability, it is possible to design one TaqMan™ probe for each allele at the SNP with dyes of different colours, so that the fluorescence ratios between both colours give a high-quality scoring of the allele. This method is simple to use and its throughput is high, but besides the rather high cost of the machine itself, double fluorescent labelling of the TaqMan™ probes is expensive and the consumable cost per sample may often be a limiting factor.

DNA chips

DNA chips are small plaques of glass or metal on which different types of single-stranded DNA molecules can be covalently bound by one extremity. Each type of DNA molecule is addressed to a small spot on the chip, in such a miniaturized way that up to one million spots can be arrayed on a 1 cm² surface. For SNP analysis, the DNA molecules spotted on the chip are 25 bp oligonucleotides designed with the sequence surrounding the SNP, the variable base being around the middle. For each allele of each SNP, the corresponding oligonucleotide is spotted at a different location of the array on the chip. PCR amplifications are performed on each individual for all the different STSs containing the SNPs, and the products are labelled (mainly by fluorescence). For each individual, the different labelled PCR products are pooled together and hybridized on the DNA chip. After the non-hybridized probes are washed, the fluorescence intensity is measured at each array location of the chip by means of a fluorescence scanner coupled with image analysis software. Practically, to ensure sufficiently high reliability of the results, five different oligonucleotides are designed for each allele of each SNP, each of them being shifted so that the position of the variable nucleotide ranges from two bases before to two bases after the middle position of the oligonucleotide. In addition, it is necessary to replicate the same oligonucleotide at two or three different locations on the chip to be able to eliminate false positives due to possible non-specific hybridization.

The use of DNA chips for SNP genotyping was reported by Wang et al. (1998) on humans, and the potential interest of this approach is constantly increasing as the density of the arrays gets higher and higher. Because of the high number of hybridizations that must be carried out simultaneously in the same conditions, the design of oligonucleotide sequences is sometimes difficult and requires particular software and expertise. Besides this, the major limitation of the technique for plant genetics is the very high cost of design of custom oligonucleotide chips. These chips are made by *in situ* synthesis of the oligonucleotides using photolithography, which requires making expensive "masters" for each set of oligonucleotides, so that the set-up cost is extremely high. However, the subsequent production of high numbers of identical chips can be cheap. The consequence is that such chips are interesting for very large projects, as in human research and diagnostics, but their field of interest for plant genetics is limited to model species or economically important crops. Other ways of using DNA chips for SNP genotyping are being investigated to overcome this limitation by designing "universal" chips that would be very cheap, but none of them is routinely applied yet. One example is given below with the OLA.

Bead-based techniques

In the DNA chips approach, ASOH is performed simultaneously on different types of oligonucleotides corresponding to each allele of each SNP, and the oligonucleotides are identified by their position on the array. Bead-based approaches are also based on simultaneous hybridizations on different oligonucleotides, but each type of oligonucleotide is identified by being bound to fluorescent micro-beads that can emit fluorescence light of a particular wavelength and intensity when excited by a laser. By combining 10 levels of intensity and two wavelengths, it is possible to distinguish between 100 different types of beads. The beads of each given type are first coated with oligonucleotides of a given sequence, corresponding to a given allele of a given SNP, and all the beads are then pooled. With 100 different types of beads, it is then possible to analyse simultaneously up to 50 SNPs (with two alleles per SNP) in a single tube.

Labelled PCR products from an individual are hybridized to the mixture of oligonucleotide-coated beads.   After washing of the non-hybridized probes, the beads are loaded into a flow cytometer, where they pass one by one through a capillary in front of a combination of two laser beams and fluorescence detectors that measure: (1) the fluorescence levels at the two wavelengths that identify the type of bead, and therefore indicate which allele of which SNP is being looked at, and (2) the fluorescence level of the PCR product, which indicates whether or not the sample hybridized. After a large number of beads have passed through the detectors (thousands in a few seconds), it is possible to calculate the mean value of PCR product fluorescence intensity for each type of bead, and to determine for each SNP which alleles are present. The potential throughput of this technique is high, but, as with DNA chips, hybridization conditions are difficult to optimize for many oligonucleotides simultaneously. The OLA technique described below provides a way of overcoming the problem of hybridization conditions. Because the method is recent, there is not much information available yet about its application in routine projects.

Primer extension

This group of methods includes variants called "minisequencing", primer extension or GOOD assay (Sauer et al., 2000). They are most of the time associated with the use of a matrix-assisted laser desorption ionization time of flight (MALDI-TOF) mass spectrometer, but primer extension reactions can also be performed in association with electrophoresis with an automatic DNA sequencer. In such methods, specific PCR is first performed with primers designed on conserved sequences to produce STS amplification with all alleles. Then, a DNA polymerization reaction is performed using the PCR product as a template, and a single primer having its 3' end immediately before the SNP.

In a first type of method, this reaction is carried out in the presence of one dideoxynucleotide triphosphate (ddNTP) and a mixture of the three other nucleotides under the normal deoxy form. The dideoxynucleotide will be incorporated if it is complementary to the matrix, but will then stop the reaction without any possible further elongation, whereas the deoxy-nucleotide allows further elongation. So the allele of the SNP that carries the base complementary to the dideoxynucleotide will produce a fragment whose size will be that of the primer plus one, whereas any of the three other possible alleles will produce a longer product, whose size will be at least that of the primer plus two, but may be longer depending on how far from the SNP is the next base complementary to the dideoxynucleotide. The size of the elongation product can then be measured accurately by MALDI-TOF mass spectrometry or electrophoresis in an automated sequencer.

The MALDI-TOF analysis is very quick (4 seconds/sample) but requires expensive equipment, high expertise, and a complete chain of robots to set up the elongation reactions fast enough to take optimal advantage of the throughput of the mass spectrometer. The approach is, therefore, better adapted for very large projects, or even multi-projects genotyping resources.

A recent variant of this technique (Sauer et al., 2002) begins with primer extension reaction with no dNTP but only ddNTPs, so that only a single-base extension occurs. The 5' part of the product is suppressed by phospho-diesterase II digestion (Fig. 8a). The molecular weight of the shortened product can then be accurately determined by MALDI-TOF mass spectrometry (Fig. 8b). The molecular weights of ddATP, ddCTP, ddGTP, and ddTTP are different enough to distinguish which ddNTP was incorporated. This technique is extremely quick and efficient in large genotyping laboratories, but requires expensive equipment and a good expertise.

In smaller laboratories, it is possible to perform single-base extension with ddNTPs fluorescently labeled, each type with a different color, so that the fluorescence color of the extension product indicates which nucleotide was incorporated. This detection is usually performed in a four-color automated sequencer. This method requires an automatic sequencer and the cost per sample is high due to fluorescent labeling, but it is efficient and simple to perform.

Pyrosequencing™

In this method, a template-dependent DNA strand elongation is performed by a DNA polymerase, starting from an oligonucleotide primer, but the dNTPs are sequentially added one by one to the reaction mix. If the dNTP added is complementary to the template, it will be incorporated to the newly synthesized strand and release a pyrophosphate group, which will be turned into ATP by an enzyme (sulphurylase) (Fig. 9a). Then the ATP transfers its energy to another enzyme (luciferase), which produces light by oxidizing its substrate (luciferin). If the dNTP added to the reaction mix is not complementary to the template, it is not incorporated, no pyrophosphate is released, and no light is emitted. After a given time following each dNTP addition, another enzyme called apyrase is added to eliminate all

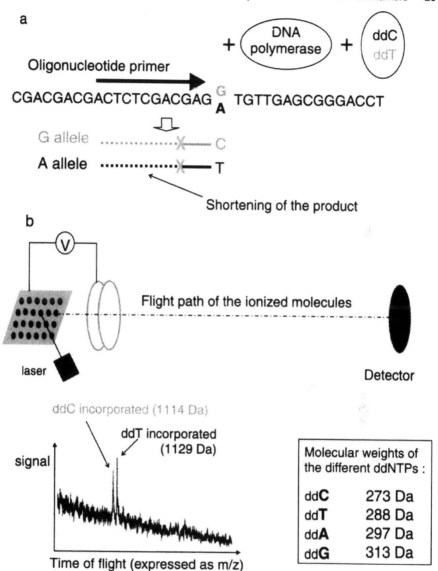

Figure 8. Principle of SNP genotyping using single-base primer extension and MALDI-TOF mass spectrometry. Depending on whether there is a C or a T at the SNP position, the molecular weight of the shortened extension product will differ by 8 base pairs, which is more than enough for them to be distinguished by MALDI-TOF. The DNA fragment mixed to a chemical (matrix) is spotted onto a plaque. Then, a laser beam ionizes the molecule and extracts it from the plaque, so that the ions are accelerated by the voltage generator (V) and conducted into the tunnel of flight. When the ion reaches the detector, the time of flight is calculated. This time is proportional to the ratio m/z, where m is the molecular weight of the ion, and z is its charge.

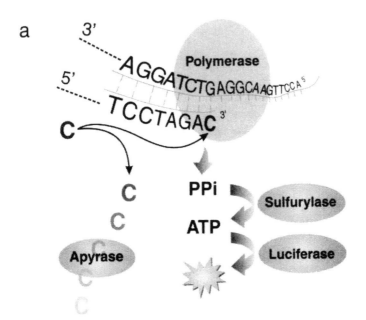

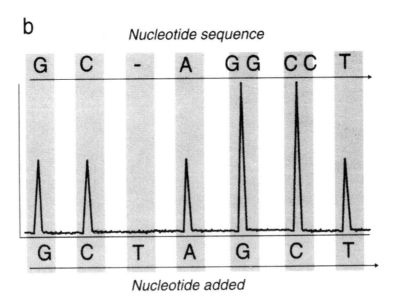

Figure 9. Principle of the pyrosequencing technique. a: The four dNTP are added one by one sequentially. When a dNTP is incorporated by the polymerase, it releases pyrophosphate (PPi), which is turned into ATP used by the luciferase to produce light. The apyrase then degrades the remaining ATP and non-incorporated nucleotides. b: By reading the peak pattern, it is easy to reconstruct the sequence of the DNA, which is here GCAGGCCT.

remaining dNTPs and ATP. A second similar cycle is then initiated by addition of the next dNTP. In such cycles, the four dNTPs are added in turn (A, G, C, T, A, G, C, T, etc.) until about 20 to 30 nucleotides have been incorporated. These reactions are carried out in an automated device that sequentially adds the different reagents and detects the light emitted by the luciferase. The intensity of light emitted after the addition of a given dNTP is proportional to the number of times this dNTP has been incorporated, which means the number of identical successive bases complementary to the dNTP in the template. The machine records the light intensity over time and indicates when each dNTP was added, so that it is possible to reconstruct the complete nucleotide sequence of the newly synthesized DNA strand (Fig. 9b). Instead of an SNP genotyping method, this technique could almost be named a sequencing method, but the length of the sequence cannot exceed about 30 bases, which limits its use for real sequencing. The technique offers a high throughput, is very simple to implement, and provides reliable information but, in addition to the cost of the machine itself, the consumables required by the analysis are expensive, which may limit the value of this approach in large-scale projects.

Oligonucleotide Ligation Assay (OLA)

An interesting property of ligases is that they cannot efficiently ligate two nucleotides if one of them makes a mismatch with its complementary strand. This provides the basics of the OLA (Baron et al., 1996): one oligonucleotide, called "capture oligo", is designed complementary to the sequence flanking the 3' side of the SNP, including the polymorphic base, and another oligonucleotide fluorescently labelled, called "reporter oligo", is designed complementary to the sequence flanking the 5' side of the SNP, excluding the polymorphic base. In the presence of the PCR product to be analysed and of a ligase, and after denaturation by heating and renaturation by cooling, the two oligonucleotides will be ligated only if the base corresponding to the SNP in the first oligo is complementary to that of the PCR product. To improve the yield of ligation, it is possible to use a thermoresistant *Taq* DNA ligase and to perform multiple thermal cycles of denaturation and ligation. This method can be used in association with DNA chips or bead-based techniques instead of ASOH, to overcome the difficult problem of designing multiple SNP-specific oligonucleotides requiring the same hybridization conditions (see for instance Iannone et al., 2000). Then, the chip location or the colors of each bead indicate which allele of which SNP is being looked at, and the fluorescence of the "reporter oligo" indicates the genotype of the individual analysed. This approach may be promising, but it is rather complex and to our knowledge is not yet completely developed as a routine tool.

Dynamic Allele-Specific Hybridization (DASH)

In this technique, the target PCR product (STS) is produced by using a biotinylated primer, so that one of the DNA strands can be bound to a streptavidin-coated microtitre plate well, and the other strand washed away with alkali. An oligonucleotide probe specific for one allele is then hybridized to the target at low temperature A double-strand–specific intercalating dye is also added to the medium, so that the dye emits fluorescence proportionate to the amount of double-stranded DNA. The sample is steadily heated while fluorescence is continually monitored. A rapid fall in fluorescence indicates the denaturing (or "melting") temperature of the probe-target duplex. When performed under appropriate conditions, a single-base mismatch between the probe and the target results in a significant lowering of melting temperature (Tm) that can be easily detected (Howell et al., 1999). This method is presented as producing reliable scoring of all SNP types and the analysis is quick with a suitable device to monitor fluorescence while heating. However, little information is yet available about its routine use.

Denaturing High-Performance Liquid Chromatography (DHPLC)

This technique uses ion-pair reversed-phase high performance liquid chromatography (HPLC) at a precisely controlled temperature, to separate DNA perfect duplexes from DNA heteroduplexes containing one or more mismatches. The column contains hydrophobic particles (stationary phase), and the mobile phase contains an ion-pairing agent that mediates the binding of DNA to the stationary phase, and an organic agent (acetonitrile) that tends to separate the DNA from the column. The column is maintained at a critical temperature, where the DNA is partly denatured, so that the heteroduplexes will elute before the homoduplexes in an increasing gradient of acetonitrile. To detect a mutation at an SNP, the PCR product to be analysed is mixed with the PCR product of a reference individual, and the mixture is denatured by heating and renatured. If the sample contains exactly the reference sequence, all duplexes will be perfect homoduplexes, and there will be only one peak of elution. If the sample is a variant at the SNP, half of the renatured molecules will be homoduplexes, and the other half heteroduplexes, so that two peaks of elution will appear. In order to obtain a clear separation between the peaks, the temperature of the column, as well as the gradient conditions, have to be set up very precisely for each particular sequence. This is made easier by specific software, and there are now specially designed HPLC systems optimized for this application. The throughput is not very high compared to other highly multiplexed techniques. However, this approach is the only one that can visualize any change in the sequence, even if its position is unknown beforehand. It can thus be used as a first step for SNP discovery: the PCR products of the different genotypes are then first compared by DHPLC to find out identical ones, and then only one of each of the different fragments has to be sequenced.

Polymorphism of number of sequence repeats: SSR

Microsatellites, or SSR (simple sequence repeats), are tandem repetitions of mono-, di-, tri-, or tetranucleotide units. The most frequent are $(A)_n$, $(TC)_n$, $(TAT)_n$, $(GATA)_n$, and so on, the value of n ranging from a few units to several tens of units (Fig. 10). Such patterns are highly abundant in eukaryotes, which means that a given SSR can be present in thousands of copies in the genome of a species (see review by Toth et al., 2000). In the higher plants, it can be estimated that there will be on the average one dinucleotide SSR for every 30 to 100 kb, with a similar density for tri- and tetra-nucleotides (Morgante and Olivieri, 1993). Such particular arrays probably originated from non-repeat sequences by random insertions or substitutions that created tandemly duplicated motifs (Zhu et al., 2000). These small repeats then become more prone to expand by particular mechanisms specific to tandem repeats, among which polymerase slippage during replication is probably the most important. This phenomenon generates frequent modifications in the number of microsatellite tandem repeats. Such sources of polymorphism are often referred to as variable number of tandem repeats (VNTR). So in addition to their distribution over the entire genome, what makes SSRs very useful in genetics is this particularly high rate of polymorphism. In soyabean, for example, up to 26 alleles of an SSR have been observed in a group of about 100 genotypes (Rongwen et al., 1995).

As the SSRs are very numerous, the Southern technique is obviously ineffective in detecting readable profiles (Fig. 10). Here again, PCR can be useful in revealing them *individually*, providing codominant and highly polymorphic locus-specific markers. Even if a given SSR is not specific to a locus, the adjacent regions are. A pair of primers specific to these adjacent regions will therefore amplify this SSR. The polymorphism in the length of the amplified SSR can be revealed either by agarose gel electrophoresis, if the differences in length between alleles are high enough, or more often in acrylamide gels (Fig. 11) and more and more frequently with automatic sequencers. SSRs constitute excellent genetic markers, with the advantages of PCR for routine detection (Fig. 12). Multiplex SSRs are an interesting illustration of this advantage (see box 1.5). SSRs have been adopted for mapping studies in maize (Senior and Heun, 1993), rice (Wu and Tanksley, 1993), barley (Sagai Maroof et al., 1994), soyabean (Morgante et al., 1994), and *Arabidopsis* (Bell and Ecker, 1994), as well as for analysis of diversity in grape (Thomas et al., 1994), rice (Yang et al., 1994), and soyabean (Powell et al., 1996). Their applications in populations have been examined by Jarne and Lagoda (1996), among others.

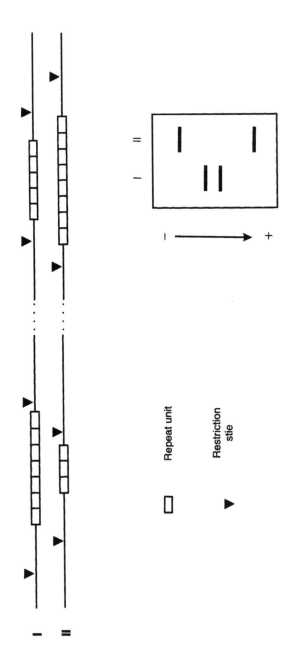

Figure 10. General structure of minisatellites and microsatellites (SSRs). In one locus, alleles differ in their number of sequence repeats, therefore in their length. If the mini- or microsatellites are used as probes in an RFLP experiment, we observe multi-band profiles, unreadable as long as the number of loci is high. The profiles observed for two loci only are represented.

The development of SSR markers is rather difficult. It is necessary to screen a genomic library (often previously enriched in SSR sequences) with an SSR probe, to sequence the positive clones, to synthesize the oligonucleotide primers, and to test the primer pairs in a sample of individuals. Different techniques for preparing SSRs in plants have been described (for example, Panaud et al., 1995; Edwards et al., 1996). For some plant species, numerous SSRs that are distributed throughout the genome have already been tested and are publicly available. Sequences of almost 2000 primer pairs of maize are thus available today on the Internet (http://www.agron.missouri.edu), and this number is constantly increasing. Around 20 SSRs have also been located in the *chloroplast* genome of *Pinus thunbergii*, the primers of which proved to be efficient in other species of pine (Vendramin et al., 1996).

PATTERNS OF MULTIPLE DOMINANT MARKERS: GENETIC FINGERPRINTING

Compared to the techniques described earlier, which all aim at revealing polymorphism within or near a particular sequence, the techniques described in this section do not target a given region of the genome determined beforehand. They reveal polymorphism *simultaneously* at several loci, which often enables the unambiguous characterization of a genotype in a single experiment. They are all not only used for fingerprinting, but also for genetic mapping, especially whenever a particular region of the genome must be densely saturated.

Sequence polymorphism

DIFFERENCES IN HYBRIDIZATION SITES OF AN ARBITRARY PRIMER, OR TECHNIQUES OF MAAP, RAPD, AP-PCR, AND DAF

The principle of MAAP consists of carrying out a PCR reaction on the DNA of the individual under study by using a primer of *arbitrary* sequence. If the primer is "not too long", and/or if the hybridization is not done in stringent conditions, the primer will hybridize to every complementary sequence in the DNA (or sequences with a limited number of mismatches). If two hybridization sites are nearby and opposite to each other, that is, in a configuration that allows PCR, the amplification will take place. If one of these two sites is absent in another individual, there will be no amplification, and a presence-absence polymorphism will be observed (Fig. 13). For a given stringency of hybridization, the shorter the primer, the higher the number of bands. The length of the primer used is moreover one of the main criteria that distinguish the three techniques based on this principle: between 5 and 15 for DAF, 9 and 10 for RAPD, and 18 and 32 for

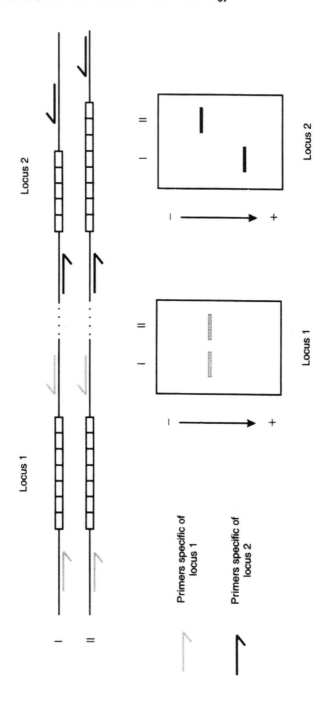

Figure 11. Detection of polymorphism of SSRs by "targeted" PCR. Using locus-specific primers, the polymorphism of different SSRs can be revealed individually.

M

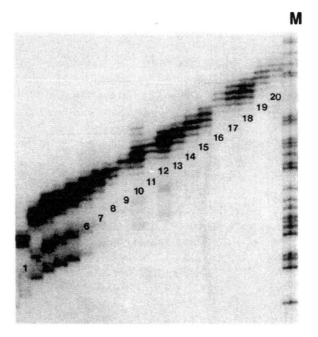

Figure 12. Example of SSR polymorphism in soyabean. It can be noted that amplification of an SSR does not produce a distinct band, but rather a group of related bands. This would be due to "slippage" of the polymerase during the PCR, causing around a major form at n sequence repeats other forms at $n \pm 1$, $n \pm 2$, $n \pm 3$, etc. (Rongwen et al., 1995).

Box 1.5: Multiplex SSR analysis

For programmes requiring the analysis of numerous markers, it is possible to visualize the polymorphism of different SSR loci on a single gel electrophoresis. A first level of multiplexing is achieved by loading together in the same gel well the PCR products of different SSRs having non-overlapping allele size ranges. For some combinations of loci, and under specially optimized PCR conditions, the different primer pairs can even be mixed within a single tube of PCR reaction, significantly reducing the cost of genotyping, in which the Taq DNA polymerase is a large part. It is also possible to load the gels several times by interrupting the run after a given migration time that will separate the different loci, so that even markers with overlapping allele size ranges can be loaded in the same gel. Finally, when automatic sequencers are used, another level of multiplexing is achieved by using different fluorescent dyes so that up to three to five different SSRs with overlapping allele size ranges can be loaded simultaneously in the same gel well or capillary. The colours of fluorescence are then detected separately to read the signals corresponding to each marker. Altogether, these different levels of multiplexing make it possible to visualize for instance the genotype of one plant at up to 16 different SSR loci in a single gel lane or capillary.

AP-PCR. Concerning this last technique, it must be noted that there is a negligible probability of finding an arbitrary sequence of 20 bases or more by chance, even in a large genome. Only when the temperature of hybridization is lowered do mismatches occur, and only then does the amplification of some fragments, even of some tens of fragments, become possible. Another difference is the migration medium. The techniques of DAF and AP-PCR may reveal very many bands (up to 100 for DAF), and acrylamide must be used (see box 1.1). On the other hand, the bands obtained in RAPD, which rarely go beyond 10, are commonly separated in agarose gel.

These techniques are not used as sources of specific markers of loci. On the one hand, we have seen, they jointly reveal *several* loci, and on the other hand, these loci are not necessarily revealed in other genetic backgrounds. There is here an important difference from the preceding techniques. The RFLP probes, or the pairs of specific primers, enable us to mark a locus in an entire species, and even in other species. Another peculiarity of RAPD, DAF, or AP-PCR is that they generally provide *dominant* markers. The reason for this is that in most cases we are dealing not with a fragment length polymorphism, but with a presence-absence polymorphism. For a given locus, fragment amplification may or may not occur. In these conditions, the homozygotes for the "band absence" allele are identified unambiguously, while the presence of a band does not enable us to distinguish between the heterozygote and the homozygote for the "band presence" allele.

When a RAPD marker has been located near a gene to be cloned, or a gene that we wish to follow through generations in a breeding project, the marker can be made specific to that particular locus. The band is excised from the gel, its DNA is sequenced, and then primers of about 20 bases are synthesized in order to specifically amplify the marker locus (these products of amplification are sometimes called sequence-characterized amplified regions, or SCAR). The various techniques for detecting polymorphism within amplified regions can thus be implemented (see box 1.4).

Techniques using arbitrary primers, especially RAPD, are popular among plant geneticists, mainly because of their simplicity. Compared to other techniques described above, they require no digestion by restriction enzyme, or transfer, or preparation and labelling of probes, or hybridization, or previous knowledge about the sequences. They are thus rapid, require little DNA, and, being based on PCR, are much more easily automated. But the results of these techniques often vary to a great extent as a response to tiny variations of experimental conditions affecting the stringency of primer annealing, so that reproducibility and transfer between laboratories have sometimes been difficult with these techniques and have limited their use. The AFLP™ technology (below) has been increasingly taking over arbitrary primer-based methods in recent

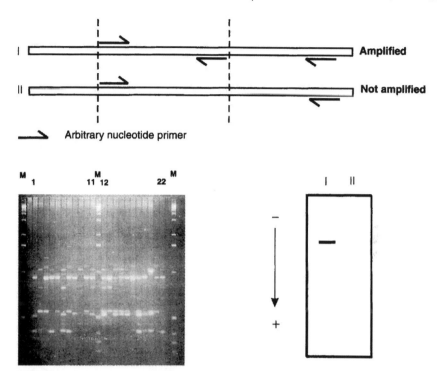

Figure 13. Polymorphism revealed by use of arbitrary primers. In individual I, two hybridization sites of primer are close together and opposite, which allows PCR amplification. One site is absent in individual II, preventing amplification. The only constraints on the design of these primers are that a certain percentage of GC must be maintained, between 45% and 65%, and that "hairpins" should be avoided. The photograph shows an example of polymorphism revealed by RAPD in a series of maize lines.

programmes, although it is not so simple to perform. As an intermediate solution, ISSR (below) is another interesting alternative. Those techniques constitute the method of choice whenever a large number of loci in a genome must be quickly marked or a particular region must be saturated. These applications are described in Chapter 3.

DIFFERENCES OF RESTRICTION SITES AND HYBRIDIZATION SITES OF SPECIFIC PRIMERS: AFLP™

For AFLP™, which is more sophisticated, the process is better controlled than with the previous techniques (Vos et al., 1995) (Fig. 14a). The DNA is first digested by two restriction enzymes with recognition sites of 6 and 4 bases such as *Eco* RI and *Mse* I, respectively. Then a ligation of cohesive ends is done with *Eco* RI and *Mse* I adapters, of known sequence, which adds 20 bases at both ends of the fragments. In a so-called pre-selection step, the fragments undergo PCR amplification using oligonucleotides as primers

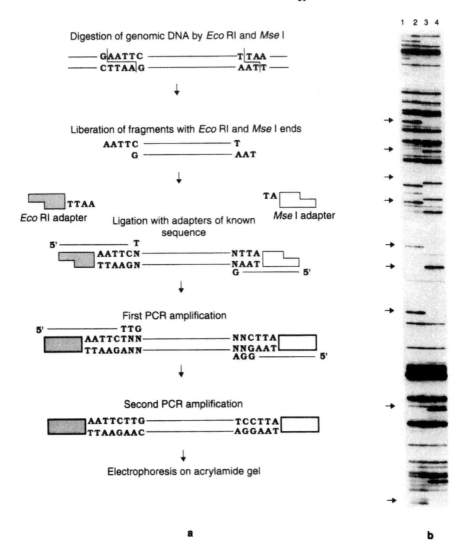

Figure 14. a: Principle of AFLP™ technique. b: Profiles of two genotypes of the ecotype Columbia (lanes 1 and 2) and of two genotypes of the ecotype Landsberg (lanes 3 and 4) of *Arabidopsis*.

corresponding to the sequence of the adapters, extended at the 3' end (towards the inside of the fragment to be amplified) by an arbitrary base. This step will thus amplify a *subset* of fragments, those that possess bases complementary to the arbitrary bases. Besides, the primers and the PCR conditions are designed to favour the amplification of *Eco* RI–*Mse* I fragments to the detriment of *Eco* RI–*Eco* RI and *Mse* I–*Mse* I fragments. Finally, a second amplification is done, using the same primers, but extended at the 3' end by two supplementary arbitrary bases. Here again, only fragments having sites complementary to arbitrary bases will be amplified. After acrylamide gel electrophoresis, several tens of fragments (up to 100) are detected, the polymorphism of which originates from restriction sites (first step) and/or hybridization sites of arbitrary bases (second and third steps) (Fig. 14b).

In studies of the plant kingdom, AFLP has been increasingly popular as a technique for genetic fingerprinting because of the high number of polymorphic bands that it provides and the quality of profiles produced by acrylamide gels. It has for instance been used for studies in population genetics in willow (Beismann et al., 1997), diversity in maize (Smith et al., 1993) and soyabean (Maughan et al., 1996b), genetic maps in barley (Becker et al., 1995), melon (Wang et al., 1997), and beetroot (Schondelmaier et al., 1996), and for identifying marker tightly linked with the R1 gene in potato (Meksem et al., 1995). It is now currently used as a source of markers for positional cloning (see chapter 3).

Modifications of the AFLP™ technique, called transposon display (Van den Broeck et al., 1998) or SSAP (sequence-specific amplification poly-morphism; Waugh et al., 1997), are based on anchoring AFLP with transposable elements, which are found in very high copy number throughout the plant genomes, sometimes preferentially in the gene-rich regions, and often account for a large proportion of the genomic DNA. In these techniques, DNA is digested and adapters ligated as for AFLP, but the PCR amplifications are performed with one primer complementary to the adapters and another one complementary to a conserved sequence of a transposable element. This approach generates genetic markers similar to AFLPs, but on average more polymorphic. Similarly, the AFLP™ technique can be modified to anchor the fragments to microsatellite sequences to take advantage of VNTR polymorphism without prior sequence knowledge.

As described for RAPD, it is possible to convert AFLP bands of interest into polymorphic STSs to produce locus-specific codominant markers easy to genotype on large populations. The AFLP bands are cut out of the AFLP gel, and the DNA is isolated, cloned, and sequenced. Specific PCR primers are then designed, and the rest of the procedure is the same as for any other STS, including the possibility of using any SNP genotyping technology (e.g. in Meksem et al., 2001).

Polymorphism of number of tandem repeats

ISSR TECHNIQUE

According to the principle of RAPD, multiple SSR loci may be revealed in the same pattern. For this, a primer made up partly of an SSR sequence and partly of arbitrary bases is used. Two types of primers are conceivable, according to the relative positions of these two parts (Fig. 15). PCR will

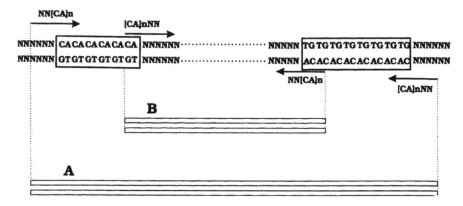

Figure 15. Principle of ISSR (or inter-SSR PCR). If the arbitrary bases of the primer are on the 5' side, one obtains the amplification product A. If they are on the 3' side, one obtains product B.

simultaneously amplify several fragments adjacent to the SSRs, which are highly polymorphic. This technique has interesting applications in plant genetics (Godwin et al, 1997), particularly for diversity and population genetics studies in plant species where no sequence information is available (e.g. Culley and Wolfe, 2001).

MINISATELLITES

Of the same general structure as microsatellites, minisatellites are distinguished by the length of the sequence repeat, which may be several tens of bases, and also by the fact that they are fewer and less uniformly distributed in the genome. They can also be used for mapping loci individually with the help of specific primers, but they are popular essentially as a source of genetic fingerprints. Jeffreys et al. (1985) showed their utility for unambiguously identifying human individuals. They are visualized by RFLP, using a restriction enzyme that cuts outside the minisatellite, and taking one part of the minisatellite itself as a probe. In principle, the number of bands revealed corresponds to the number of minisatellite loci in the genome (Fig. 16). In practice, one can only correctly read the regions of the migration lane that are not too crowded. Moreover, the markers are

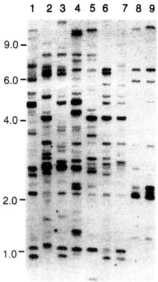

Figure 16. Example of RFLP profiles obtained for 9 rice varieties using a minisatellite probe. The scale of molecular weights is in kilobases (after Dallas, 1988).

dominant because the genetic analysis most often is inadequate to identify the allelic bands. In plants, minisatellites have rapidly been supplanted by RAPD and other PCR-based techniques.

Gene markers: cDNA and proteins

The term "marker" often implies that the loci detected are anonymous and phenotypically neutral. Even if the marker corresponds to a known-function gene, which is rarely the case, it is not useful because of its function, but because it tags a given portion of the genome. For numerous applications of markers, such as construction of genetic maps, positional cloning, or studies of diversity, the *nature* of markers is not important.

The construction of *gene* maps is nevertheless a major goal for the study of genome organization, comparative mapping, knowledge of gene duplication and distribution, as sources of candidates for characterization of quantitative trait loci, etc. Genes that are already cloned and cDNA of known function constitute the main source of gene markers, whether they are used as RFLP probes or as STS. Their genetic mapping is in progress in some agronomically important species such as maize, barley, *Brassica* or tomato, where programmes of expressed sequence tag (EST; see glossary) isolation have been initiated. Of course complete genome sequencing provides exhaustive information, but so far it is restricted to the model species with small genome size, *Arabidopsis* and rice.

Isozymes have already made it possible to map 20 or 30 genes in some species, but two-dimensional protein electrophoresis may be a useful secondary technique. Indeed, cDNA mapping reveals sometimes several loci (on the average 1.7 in maize), of which it is difficult to say whether they correspond to pseudogenes or to multigene families, and which of them are expressed. Two-dimensional protein electrophoresis is a source of *expressed* genome markers (transcribed and translated). Certain proteins present variations in position, and genetic analysis of them shows that they are monogenic and codominant (de Vienne et al., 1996). These variations are due, in most cases, to a polymorphism of the gene that encodes the protein (Touzet et al., 1995a). The mapping of the loci responsible for these variations enables us to construct a map of genes expressed in the organ analysed.

What markers are suitable for what purpose?

The general characteristics of the major types of markers are summarized in Table 5. These have of course only an indicative value. In genetic fingerprinting techniques, we can sometimes find allelic bands, and RFLP always provides some dominant loci. Moreover, as we have seen, we can transform markers from RAPD or AFLP™ into single-locus codominant markers. But the great division proposed, based on the genetic criterion, responds to different needs.

Table 5. Comparing the most common markers

Markers	Neutrality	No.	Codominance	Locus specificity[1]	Polymorphism	Organ-stage[2]	Technical complexity	Coding sequence
Morphological	No	Limited·	Rare	Yes	Low	Yes	Low	–
Isozymes	Yes	≤ 40	Yes	Yes	Low	Yes	Low	Yes
Proteins (2D)	Yes	≤ 100	Yes	Yes	Low	Yes	Medium	Yes
RFLP	Yes	$\approx$ unlimited	Yes	Yes	High	No	High	Yes or no
SSR	Yes	Thousands	Yes	Yes	Very high	No	High[3]	No
MAAP	Yes	$\approx$ unlimited	No	No	Very high	No	Low	Yes or no
ISSR	Yes	$\approx$ unlimited	Not evaluated	No	Very high	No	Low	No
AFLP	Yes	$\approx$ unlimited	No	No	Very high	No	Medium	Yes or no
STS	Yes	Limited	Yes	Yes	Moderate	No	High[3]	Yes
SNP	Yes or no	Unlimited	Yes	Yes	Moderate	No	High	Yes or no

1. "Yes" is for the techniques that reveal the marker loci individually (or in very small numbers); "No" those that reveal patterns of multiple loci.
2. Markers can be revealed only in some organs or developmental stages.
3. For SSR and STS, development of primers for a large number of informative marker loci is technically difficult and costly, but afterward routine analyses can be done much more quickly than with RFLP.

Genetic fingerprinting techniques are suitable in three situations:
- Studies of diversity: e.g., intra- or interpopulation structure, genetic distances, clustering.
- Positional cloning: they enable us to obtain a high density of markers in a particular genome region, near the gene to be cloned.
- Rapid saturation of genetic maps, in species that are seldom studied, where probes or PCR markers have not been developed.

Codominant markers revealed individually, on the other hand, are used for:
- Construction of consensus maps of a species, and whenever markers must be exchanged between laboratories.
- Obtaining better precision in the maps when the populations being mapped include heterozygous individuals.
- Comparative mapping between species.
- Mapping of quantitative trait loci when their dominance must be measured.
- Construction of gene maps.
- Population genetics studies.

Whatever the application considered, the choice of a technique will also depend on the species studied. Microsatellites or AFLP, for example, are sources of markers that are very useful for species that have low levels of polymorphism. On the other hand, for a species having a high level of polymorphism and many cDNA, genes or anonymous probes that are already cloned, RFLP, or better, STS, are adapted to a large variety of applications. In the Gymnosperms, where the haploid megagametophytes offer the possibility of constructing a genetic map per individual, techniques of the MAAP or STS type are recommended rather than RFLP, which requires quantities of DNA incompatible with the size of the organ and is too difficult for analysis of a large number of individuals. These different applications will be explained in detail later in the book.

Glossary

AFLP™: Amplified Fragment Length Polymorphism

PCR amplification of genome DNA after digestion with two restriction enzymes and ligation of an adapter of around 20 base pairs. The primers are made up of the adapter plus 3 random bases added at the 3' end.

Vos, P., Hogers, R., Bleeker, M., Reijans, M., van de Lee, T., Hornes, M., Frijters, A., Pot, J., Peleman, J., Kupier, M., and Zabeau, M. 1995. AFLP: a new technique for DNA fingerprinting. *Nucl. Acids Res.*, 23, 4407-4414.

AP-PCR: Arbitrarily Primed PCR

PCR amplification of unknown sequences using a single primer of 20 arbitrary bases, with low temperatures of hybridization.

Welsh, J., and McClelland, M. 1990. Fingerprinting genomes using PCR with arbitrary primers. *Nucl. Acids Res.*, 19, 861-866.

ASOH: Allele-Specific Oligonucleotide Hybridization

Hybridization of a PCR product containing an SNP with an oligonucleotide that will only hybridize to one allele of the SNP.

Conner, B.J., Reyes, A.A., Morin, C., Itakura, K., Teplitz, R.L., and Wallace, R.B. 1983. Detection of sickle cell β^S-globin allele by hybridization with synthetic oligonucleotides. *Proc. Natl. Acad. Sci. USA*, 80, 278-282.

CAPS: Cleaved Amplified Polymorphic Sequence

An amplified genome fragment (STS or SCAR) is digested by restriction enzymes, then the length polymorphism generated is revealed by electrophoresis. Other name: PCR-RFLP.

Konieczny, A., and Ausubel, F.M. 1993. A procedure for mapping *Arabidopsis* mutations using codominant ecotype-specific PCR-based markers. *Plant J.*, 4, 403-410.

DAF: DNA Amplification Fingerprinting

Same principle as RAPD using shorter primers (5 to 10 bases). The electrophoretic profiles of amplified DNA fragments are often highly complex.

Caetano-Anolles, G., Bassam, B.J., and Gresshoff, P.M. 1991. DNA fingerprinting using very short arbitrary oligonucleotides. *Bio/ technology*, 9, 553-557.

DASH: Dynamic Allele-Specific Hybridization

SNP genotyping technique based on the precise measure of DNA melting temperature by monitoring the amount of double-strand DNA during heating.

Howell, W.M., Jobs, M., Gyllensten, U., and Brooks, A.J. 1999. Dynamic allele-specific hybridization: A new method for scoring single nucleotide polymorphisms. *Nat. Biotechnol.*, 17, 87-88.

DGGE: Denaturing Gradient Gel Electrophoresis

Electrophoresis in acrylamide gel containing a linear gradient of urea/ formamide. The DNA fragment, initially double-stranded, is partly denatured to form a "branched" structure (Y-shaped), with almost no mobility. Differences in sequences in the least stable melting region lead to differences of stability, and therefore of distance of migration. The principle of TGGE (temperature gradient gel electrophoresis) is the same, except that a temperature gradient is used instead of a urea/ formamide gradient.

Myers, R.M., Maniatis, T., and Lerman, S. 1987. Detection and localization of single base changes by denaturing gradient gel electrophoresis. *Methods Enzymol.*, 155, 501-527.

DHPLC: Denaturing High Performance Liquid Chromatography

SNP genotyping technique based on the detection of mismatches in double strand DNA by column chromatography.

Giordano, M., Oefner, P.J., Underhill, P.A., Cavalli Sforza, L.L., Tosi, R., and Richiardi, P.M. 1999. Identification by denaturing high-performance liquid chromatography of numerous polymorphisms in a candidate region for multiple sclerosis susceptibility. *Genomics*, 15, 247-253.

EST: Expressed Sequence Tags

Anonymous sequences or sequences of unknown function, arising from systematic programmes of cDNA sequencing, enabling identification of clones never isolated in the species being studied, and also construction of gene maps.

Adams, M.D., Kelley, J.M., Gocayne, J.D., Dubnick, M., Polymero-poulos, M.H., Xiao, H., Merril, C.R., Wu, A., Olde, B., Moreno, F., Kerlaage, A.R., McCombie, R.W., and Venter, J.C. 1991. Complementary DNA sequencing: expressed sequence tags and human genome project. *Science*, 252, 1651-1656.

GOOD Assay

Optimization of the primer extension-MALDI-TOF analysis for efficient SNP genotyping.

Sauer, S., Lechner, D., Berlin, K., Lehrach, H., Escary, J.L., Fox, N., and Gut, I.G., 2000. A novel procedure for efficient genotyping of single nucleotide polymorphisms. *Nucl. Acids Res.*, 28, e13.

IDP: Insertion-Deletion Polymorphism

Polymorphism caused by the insertion or deletion of a DNA sequence. The polymorphism of VNTRs is a particularly polymorphic type of IDP.

ISSR: Inter Simple Sequence Repeats

PCR amplification on genomic DNA with primers that carry dinucleotide repeated patterns and some arbitrary bases. The amplified fragments are thus located between microsatellite loci. Other names: IMA, Inter SSR PCR, ISA, IRA, RAMP.

Zietkiewicz, E., Rafalski, A., and Labuda, D. 1994. Genome fingerprinting by simple-sequence repeat (SSR)-anchored polymerase chain reaction amplification. *Genomics*, 20, 176-183.

MAAP: Multiple Arbitrary Amplicon Profiling

Generic term denoting all the PCR amplification techniques on genomic DNA that use arbitrary primers and generate complex and polymorphic electrophoretic profiles (RAPD, AP-PCR, DAF, AFLP™, IMA).

Caetano-Anolles, G. 1994. MAAP: a versatile and universal tool for genome analysis. *Plant Mol. Biol.*, 25, 1011-1026.

MALDI-TOF: Matrix-Assisted Laser Desorption-Ionization Time of Flight

Mass spectrometer capable of measuring very accurately the molecular weight of small DNA fragments, used for SNP genotyping.

Sauer, S., Lechner, D., Berlin, K., Lehrach, H., Escary, J.L., Fox, N., and Gut, I.G. 2000. A novel procedure for efficient genotyping of single nucleotide polymorphisms. *Nucl. Acids Res.*, 28, e13.

Microsatellite

Repetitions in tandem of mono-, di-, tri-, or tetranucleotide units. Examples: $(A)_n$, $(TC)_n$, $(TAT)_n$, $(GATA)_n$. Such motifs are abundant and highly polymorphic in the genome of eukaryotes. Also called SSR (simple sequence repeat).

Tautz, D. 1989. Hypervariability of simple sequences as a general source for polymorphic DNA markers. *Nucl. Acids Res.*, 17, 6463-6471.

OLA: Oligonucleotide Ligation Assay

SNP genotyping techniques based on a ligation between two oligonucleotides.

Baron, H., Fung, S., Aydin, A., Bahring, S., Luft, F.C., and Schuster, H. 1996. Oligonucleotide ligation assay (OLA) for the diagnosis of familial hypercholesterolemia. *Nat. Biotechnol.*, 14(10), 1279-1282.

PCR: Polymerase Chain Reaciton

Enzymatic amplification *in vitro* of a DNA sequence from a pair of specific oligonucleotide primers. The enzyme used is a thermostable DNA polymerase (Taq DNA polymerase).

Saiki, R.K., Gelfland, D.H., Stoffel, S., Scharf, S.J., Higuchi, R., Horn, G.T., Mullis, K.B., and Ehrlich, H.A. 1988. Primer-directed enzymatic amplification of DNA with thermostable DNA polymerase. *Science*, 239, 487-491.

PCR-RFLP
See CAPS.
RAPD: Random Amplified Polymorphic DNA

PCR amplification of DNA fragments using a primer of 10 arbitrary bases.

Williams, J.G.K., Kubelik, A.R., Livak, K.J., Rafalski, J.A., and Tingey, S.V. 1990. DNA polymorphisms amplified by arbitrary primers are useful as genetic markers. *Nucl. Acids Res.*, 18, 6531-6535.

RFLP: Restriction Fragment Length Polymorphism

Polymorphism generated by hydrolysis of DNA by a restriction endonuclease. The polymorphism is revealed after hybridization of a labelled probe on the fragments separated by gel electrophoresis and transferred on a membrane (Southern technique).

Botstein, D., White, R.L., Skolnick, M., and Davis, R.W. 1980. Construction of a genetic linkage map in man using restriction fragment length polymorphism. *Am. J. Human Genet.*, 32, 314-331.

Primer Extension

In vitro DNA polymerization reaction in the presence of various combinations of dNTPs and/or ddNTPs. This technique is often used for SNP genotyping, coupled with MALDI-TOF mass spectrometry.

SCAR: Sequence-Characterized Amplified Region

Genomic DNA fragment amplified by PCR with specific primers (14 to 20 bases). These primers are defined from the sequence of an RAPD fragment of interest excised from a gel, then cloned and sequenced.

Paran, I., and Michelmore, R.W. 1993. Development of reliable PCR-based markers linked to downy mildew resistant genes in lettuce. *Theor. Appl. Genet.*, 85, 985-993.

SNP: Single Nucleotide Polymorphism

Nucleotide site for which substitution polymorphism has been observed at a significant frequency between different individuals.

Wang, D.G., Fan, J.B., Siao, C.J., Berno, A., Young, P., Sapolsky, R., Ghandour, G., Perkins, N., Winchester, E., Spencer, J., Kruglyak, L., Stein, L., Hsie, L., Topaloglou, T., Hubbell, E., Robinson, E., Mittmann, M., Morris, M.S., Shen, N., Kilburn, D., Rioux, J., Nusbaum, C., Rozen, S., Hudson, T.J., Lipshutz, R., Chee, M., and Lander, E.S. 1998. Large-scale identification, mapping, and genotyping of single-nucleotide polymorphisms in the human genome. *Science*, 280, 1077-1082.

SSAP (or S-SAP): Sequence-Specific Amplification Polymorphism

Fingerprinting method derived from AFLP™, which amplifies fragments containing sequences of retrotransposable elements.

Waugh, R., McLean, K., Flavell, A.J., Pearce, S.R., Kumar, A., Thomas, B.B.T., and Powell, W., 1997. Genetic distribution of Bare-1-like retrotransposable elements in the barley genome revealed by sequence-specific amplification polymorphism (S-SAP). *Mol. Gen. Genet.*, 253, 687-694.

SSCP: Single Strand Conformation Profile (or Polymorphism)

Conformation polymorphism of single-stranded DNA revealed in a non-denaturing acrylamide gel. Before gel loading, the double-stranded DNA fragments are denatured, and the solution is quenched in order to prevent their reassociation into a double strand.

Orita, M., Iwahana, H., Kanasawa, H., Hayashi, K., and Sekiya, T. 1989. Detection of polymorphism of human DNA by gel electrophoresis as single-strand conformation polymorphism. *Proc. Natl. Acad. Sci. USA*, 86, 2766-2770.

SSR: Simple Sequence Repeats
See Microsatellite.
STS. Sequence Tagged Site

PCR amplification of a sequence with specific primers. Sometimes used to define the conversion of an RFLP probe into locus-specific PCR test.

Olson, M., Hood, L., Cantor, C., and Doststein, D. 1989. A common language for physical mapping of the human genome. *Science*, 254, 1434-1435.

TGGE: Temperature Gradient Gel Electrophoresis
See DGGE.
Tm: Melting Temperature

Temperature at which statistically half of the DNA molecules are in the denatured form. This temperature depends on the DNA fragment length and sequence, and on the composition of the buffer. For oligonucleotides, an approximate value of Tm is given by: 2 × (number of A or T bases) + 4 × (number of G or C bases).

VNTR: Variable Number Tandem Repeats

Generic term denoting microsatellites and minisatellites, which are prone to very frequent insertion-deletion polymorphism of repeat units.

Hahn, M., Serth, J., Fislage, R., Wolfes, H., Allhoff, E., Jonas, V., and Pingoud, A. 1993. Polymerase chain reaction detection of a highly polymorphic VNTR segment in intron 1 of the human p53 gene. *Clin. Chem.*, 39, 549-550.

2

Construction of Genetic Linkage Maps

D. de Vienne

The concept of genetic linkage has been known since the studies of Morgan (1911). As early as 1913, Sturtevant published a genetic map of the chromosome X of Drosophila, and the first partial genetic map of maize was published in 1935 by Emerson et al. When morphological markers are used, the number of loci segregating in a given population remains limited. The maps of morphological markers have been constructed from data compiled from various experiments. With biochemical markers, and above all molecular markers, it is possible to *jointly* map a large number of loci within a single population, until the genome is "saturated", that is, until every point of the genome is genetically linked to at least one marker. In such maps, there are thus as many linkage groups as there are chromosomes.

In plants, the conditions are generally favourable for the construction of genetic maps. Controlled crosses can be done, homozygous parents are often available or can be developed, and large populations can be used. Still, it is possible to use populations derived from non-fixed parents, as in strictly allogamous species, as we will see.

The most commonly used populations are generally derived from two homozygous parents crossed to produce F_1 individuals, all identical and heterozygous for all the loci that have fixed different alleles in the parents (Fig. 17) (we must therefore choose, as far as possible, sufficiently different parents so that the number of polymorphic loci is not limiting). Three major types of mapping populations can be derived from such a cross: those that give access to the haploid phase, F_2 populations, and populations of recombinant inbred lines (only diploid species will be considered hereafter). Before describing them in detail, we will present the fundamentals of the concept of genetic distance.

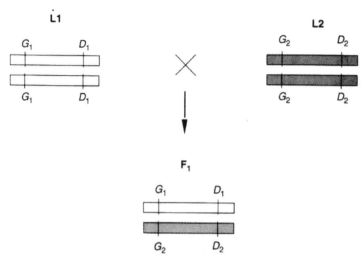

Figure 17. Initial crossing between inbred lines at the origin of the most common mapping populations. *G* and *D* represent any two loci between which a genetic distance is to be calculated. G_1 and G_2 (or D_1 and D_2) symbolize two alleles at locus *G* (or *D*). If locus *G* (or *D*) is revealed by a molecular marker technique, then alleles G_1 and G_2 (or D_1 and D_2) are revealed for example by the two bands of a microsatellite profile (codominant marker), or by the presence or absence of a band in RAPD or AFLP (dominant marker).

THE CONCEPT OF GENETIC DISTANCE

Let us imagine that we have access to the genotype of gametes resulting from the meiosis of F_1 individuals (Fig. 18) (in plants, the gametes do not directly result from meiosis, since there is the gametophytic phase between the two; but that does not alter the genetic argument). If two distinct loci *G* and *D* are considered, 4 different haploid gametes can be formed, which are either of the "parental" type or of the "recombinant" type. The proportion of recombined gametes, or recombination rate, *r*, is related to the distance between the loci (the longer this distance, the greater the probability of crossover). The construction of a genetic linkage map involves estimation of recombination rates between all the loci, two by two. However, *r* cannot be used as such as a measure of the distance between two loci. Its minimum value is 0, and its maximum value is limited to 0.5, which corresponds to the situation of genetic independence of the loci. The recombination rates are therefore not additive. There are two reasons for this:

- Multiple crossovers. As the distance between two loci increases, the probability of more than one crossover between them increases. The rate of recombination therefore underestimates the genetic distance, because the effects of two crossovers are cancelled out (the argument extends, of course, to higher even numbers of crossovers, even though this situation is practically improbable).

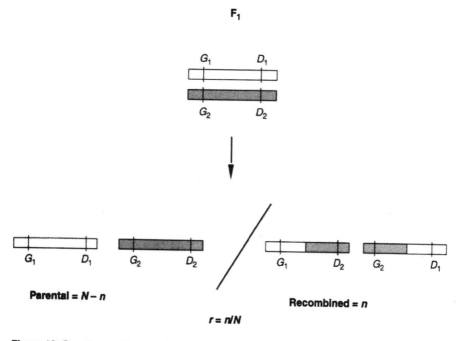

Figure 18. Genotypes of the possible gametes (or gametophytes) created by the meiosis of an F_1 individual, for loci G and D. From the recombination rate, $r = n/N$ (where N is the total number of gametes), the genetic distance between these two loci can be calculated.

- Independent of probabilistic aspects, two crossovers can hardly occur close to each other. This is the phenomenon of *interference*, which is supposed to originate from mechanical constraints during meiosis.

It is therefore necessary to define mapping functions that are additive, i.e. mapping functions that account for the number of crossovers from the estimation of r.

The Haldane distance

A distance considering only multiple crossovers was defined by Haldane in 1919 and can be introduced in two ways:

1) Given three genetically linked loci A, B and C:

A recombination can be observed between A and C if there has been recombination between A and B *and* if there has not been any between B and

C, or if there has been recombination between B and C *and* if there has not been any between A and B. We can therefore write:

$$r_{AC} = r_{AB}(1 - r_{BC}) + r_{BC}(1 - r_{AB})$$

or

$$r_{AC} = r_{AB} + r_{BC} - 2r_{AB}r_{BC},$$

where r_{AB}, r_{BC} and r_{AC} are the recombination rates between A and B, B and C, and A and C, respectively. This equation shows that the recombination rates are clearly non-additive.

This equation can be put in the following form:

$$(1 - 2r_{AC}) = (1 - 2r_{AB})(1 - 2r_{BC})$$

and may be generalized to any number of loci.   A logarithmic transformation makes the relation linear:

$$\ln(1 - 2r_{AC}) = \ln(1 - 2r_{AB}) + \ln(1 - 2r_{BC})$$

A function of r in the form:

$$m = p \ln(1 - 2r),$$

where p is a constant, constitutes an additive distance. As one must have $m \approx r$ for the very small values of r, and as $\ln(1 - \varepsilon) \approx -\varepsilon$, we must choose $p = -1/2$, from which we get the distance:

$$m = -(1/2) \ln(1 - 2r)$$

This so-called Haldane distance (1919) is expressed in morgans, but it is classically expressed in centimorgans (cM), i.e., it is written as follows:

$$m_H = -50 \ln(1 - 2r)$$

2) Given X, a random variable associated with the number of crossovers between two loci. This variable follows a Poisson's law of expectation m:

$$P(X = k) = e^{-m}(m^k / k!)$$

Recombination between the two loci cannot be observed if there is no crossover or if there is an even number of crossovers. The probability of *not* observing recombination is therefore:

$$P(X = 2k) = (1 - r) = \sum_{k=0}^{\infty} e^{-m}(m^{2k}/(2k)!)$$

Since

$$e^m = 1 + m + m^2/2! + m^3/3! + m^4/4! + \dots \text{ etc.}$$

and

$$e^{-m} = 1 - m + m^2/2! - m^3/3! + m^4/4! - \ldots \text{ etc.,}$$

we have

$$(e^m + e^{-m})/2 = \sum_{k=0}^{\infty} m^{2k}/(2k)!$$

and thus

$$(1 - r) = e^{-m}(e^m + e^{-m})/2$$

Hence

$$m = -(1/2) \ln(1 - 2r)$$

that is the Haldane distance.

The Kosambi distance and other distances

There is a more general formulation of genetic distance, which includes a parameter that takes into account the possible occurrence of interference. Let us take the same three linked loci A, B and C, but supposing that there is interference. In this case the probability of crossover in an interval is reduced by the existence of a crossover in the adjacent interval. The rate of reduction, c, called *coincidence*, is the ratio of the actual number of crossovers to the number expected in the absence of interference. Taking again the preceding argument, we can write:

$$r_{AC} = r_{AB}(1 - cr_{BC}) + r_{BC}(1 - cr_{AB})$$

or

$$r_{AC} = r_{AB} + r_{BC} - 2cr_{AB}r_{BC}$$

Let us call the difference between distances AB and AC Δm, and let us suppose the distance BC is very small. We then have $r_{BC} \approx \Delta m$, from which

$$\Delta r = r_{AC} - r_{AB} = \Delta m(1 - 2cr_{AB})$$

And proceeding to the limit

$$dm/dr = 1/(1 - 2cr)$$

If r is very small, whether or not there is interference, we have $dm/dr \approx 1$, and we find therefore that $r \approx m$.

For any r, and in the absence of interference, that is for $c = 1$, we have

$$m = \int_{r=0}^{\infty} dr / (1 - 2r)$$

from which

$$m = -(1/2) \ln(1 - 2r)$$

where we again find the Haldane distance.

In the presence of interference, we must attribute a value to c. The value of c is proportionate to r, but no universal relation between c and r can be given, because the magnitude of interference varies according to the species, chromosomes, and possibly chromosomal regions. Kosambi (1944) proposed to take $c = 2r$, which leads to $c = 0$ when $r = 0$, and $c = 1$ when $r = 0.5$.

We thus obtain

$$m = \int_{r=0}^{\infty} dr / (1 - 4r^2)$$

from which

$$m = (1/4) \ln[(1 + 2r)/(1 - 2r)]$$

Here again, to express the distance in cM, we take

$$m_K = 25 \ln[(1 + 2r)/(1 - 2r)]$$

which is called the Kosambi distance.

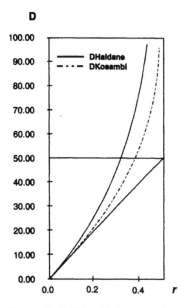

Figure 19. Relationship between the recombination rate r and the Haldane and Kosambi distances (expressed in cM).

Figure 19 represents the variation of Haldane and Kosambi distances as a function of r. The lowest value of Kosambi's distance accounts for the phenomenon of interference. These two distances are by far the most often used in plant genetics, but there are others, which differ in the relation between c and r. One such distance uses in particular the relation $c = (2r)^2$, from which we always have $c = 0$ when $r = 0$ and $c = 1$ when $r = 0.5$, and which would adjust well to data of certain species (Crow, 1990). But beyond these empirical distances, it must be emphasized that the validity of a mapping function can and must be verified afterwards, in order to ensure that the distances obtained are actually additive. Moreover, we must keep in mind that a given distance may be valid in some genome regions and not in others.

These theoretical developments enable us to specify the concept of centimorgan. Contrary to what is often read and taught, one centimorgan *does not correspond* to a recombination rate of 1%. Even if these quantities are very close, writing the "equal" sign between them would imply, for example, that 30 cM would correspond to 30% of recombination, which is obviously wrong (Fig. 19). In this connection, it is worth noting that two loci 50 cM apart *are genetically linked* ($r \approx 0.32$). In reality, the centimorgan is the distance separating two loci between which *the expectation of number of crossover is one hundredth* (0.01).

Estimation of recombination rates and linkage tests

Even though the recombination rate r is defined as the ratio of the number of recombinant gametes to the total number of gametes, and though we do not have access to the genotype of the gametes, r can be estimated directly in certain populations (see below). In other cases, especially in F_2 populations, it is necessary to use the maximum likelihood method to estimate its most probable value (see box, 2.1). Finally, with the recombinant inbred lines, a formula must be used to deduce r from the number of lines that have recombined (see below).

A statistical test is needed before declaring linkage between two loci. In practice, the χ^2 is sometimes used (see box, 2.2), but most often the LOD score, derived from methods of maximum likelihood, is used (box, 2.3). Gerber and Rodolphe have analysed the relations between these two tests (1994a). Once the distances between all the loci, two by two, have been estimated, the loci can be organized into linkage groups. Three-point and multi-point tests are then used to specify their order (see box, 2.4). When the genome is saturated with markers, the distances may be added to calculate its total *genetic* length (Fig. 20). More details on genetic linkage can be found in Bailey (1961).

COMPARING THE MOST COMMONLY USED POPULATIONS

Populations giving access to the haploid phase

The case in which one has access to the haploid phase, described above, is not only a theoretical one. Even in the higher Eukaryotes, there are several situations in which, at least formally, the haploid case applies exactly.

DOUBLED HAPLOIDS

In various species (among Poaceae, Solanaceae, etc.), viable plants can be regenerated from microspores (androgenesis) or macrospores (gynogenesis).

Box 2.1: Estimating the recombination rate by the maximum likelihood method

The basic idea of calculating the maximum likelihood is to consider that the result observed, that is the sizes n_i of the different genotype classes of the population, was the most probable given the value of the parameter to be estimated (r). We therefore look for the value of r that maximizes the probability of the distribution of n_r. The multinomial law gives this probability:

$$P(X_1 = n_1, ..., X_t = n_t) = L = \frac{N!}{\prod_i n_i!} \prod_t m_i^{n_j}$$

where N is the size of the progeny, m_i is the theoretical frequency of the genotypic class i (function of r), and t is the total number of classes. We have $t = 4$ in the case of a backcross; in the case of an F_2, we have $t = 9$ if both loci are codominant, $t = 6$ if one locus is dominant and the other is codominant, and $t = 4$ if both loci are dominant.

To find the maximum of this function of r, designated L for likelihood, simply find the value of r equating its derivative to zero or, what is the same, the value equating the derivative of the logarithm to zero (the advantage of using logarithms is that the products become sums and the calculation is made possible).

Where we have access to the haploid phase (doubled haploids, backcross, gametophytes), we have:

$$L = \frac{N!}{\prod_i n_i!} \left[(1-r)/2\right]^{n_1+n_4} (r/2)^{n_2+n_3}$$

where n_1 and n_4 are the numbers of parental genotypes, and n_2 and n_3 are those of the recombined genotypes. It is easily shown that the likelihood function $d\ln L/dr$ equals zero for $r = (n_2 + n_3)/N$, which corresponds to the very definition of the rate of recombination between two loci.

The case of F_2 populations is not so straightforward. If there is codominance at both loci, we have:

$$L = \frac{N!}{\prod_i n_i!} \left[(1-r)^2/4\right]^{n_1+n_2} \left[r(1-r)/2\right]^{n_3+n_4+n_6+n_7} (r^2/4)^{n_8+n_9} \left[(1-r)^2/2+r^2/2\right]^{n_5}$$

where the quantities between square brackets are the frequencies from Table 7, and the n_i are the sizes of different genotype classes.

The condition $d\ln L/dr = 0$ becomes: $\alpha_1 r^3 - \alpha_2 r^2 + \alpha_3 r - \alpha_4 = 0$

where: $\alpha_1 = 4N$

$\alpha_2 = 4(n_1 + n_2) + 6(n_3 + n_4 + n_5 + n_6 + n_7) + 8(n_8 + n_9)$

$\alpha_3 = 2(n_1 + n_2 + n_5) + 4(n_3 + n_4 + n_6 + n_7) + 6(n_8 + n_9)$

$\alpha_4 = n_3 + n_4 + n_6 + n_7 + 2(n_8 + n_9)$

This equation of the third degree has a root between 0 and 0.5 that can be found by iteration.

Whenever there is codominance for one locus and dominance for the other, or dominance for both loci, the likelihood function must be recalculated, but the principle is exactly the same. Mapping software exist for these calculations.

Box 2.2: Test of linkage between two loci: χ^2 test

In the case of the haploid phase (doubled haploids, backcross, gametophytes), in the absence of genetic linkage and segregation distortion, the theoretical and observed values of the four possible genotypes are:

Genotypes	G_1D_1	G_1D_2	G_2D_1	G_2D_2
Theoretical values	$N/4$	$N/4$	$N/4$	$N/4$
Observed values	n_1	n_2	n_3	n_4

where N is the size of the population.

A χ^2 test with 3 df for adjustment of data to theoretical values can be done, but the conclusion will be ambiguous in case of significant result. Indeed, the deviations from theoretical values can be due not only to genetic linkage, but also to distortions of segregation at one locus or the other. This χ^2 can be partitioned into a sum of 3 χ^2 with 1 df, 2 to test the distortions at the loci, and 1 to test the linkage. For the distortion of segregation at locus G, we must test (χ^2_G):

$$n_1 + n_2 = n_3 + n_4 \text{ against } n_1 + n_2 \neq n_3 + n_4$$

for the distortion at locus D (χ^2_D):

$$n_1 + n_3 = n_2 + n_4 \text{ against } n_1 + n_3 \neq n_2 + n_4$$

and finally for the linkage (χ^2_L):

$$n_1 + n_4 = n_2 + n_3 \text{ against } n_1 + n_4 \neq n_2 + n_3$$

We can then easily calculate the values of χ^2:

$$\chi^2_G = (n_1 + n_2 - n_3 - n_4)^2/N$$

$$\chi^2_D = (n_1 - n_2 + n_3 - n_4)^2/N$$

$$\chi^2_L = (n_1 - n_2 - n_3 + n_4)^2/N$$

It must be noted that in case of distortion at both loci (and not at only one), the χ^2_L of the linkage can be found significant even in the absence of the linkage. Indeed, in this case $n_1 + n_4$ cannot be equal to $n_2 + n_3$. If the distortion at the two loci is in the same direction (the same parent carries the over- or under-represented alleles), a pseudo-linkage will be observed. If on the other hand the over- or under-represented alleles come from different parents, the observed recombination rate will be higher than 0.5. The χ^2_L is therefore not suitable here, and we must use an adjustment χ^2_L on the following data:

Genotypes	G_1D_1	G_1D_2	G_2D_1	G_2D_2
Theoretical values	psN	$p(1-s)N$	$(1-p)sN$	$(1-p)(1-s)N$
Observed values	n_1	n_2	n_3	n_4

where p is the estimated frequency of allele G_1 and s is the estimated frequency of allele D_1. The theoretical numbers are simply calculated from the products of allele frequency at the two loci, since the null hypothesis is absence of linkage. If $p = 0.5$ and $s = 0.5$, we find well $N/4$.

Box 2.3: Test of linkage between two loci by the LOD score method

The most common method for testing a linkage between two loci (the one commonly used in mapping software) is calculation of a likelihood ratio, or "LOD score" (logarithm of the odds ratio):

$$LOD = \log_{10} [L(r)/L(r_0)]$$

where $L(r)$ is the likelihood calculated with the most probable value of r (or maximum likelihood, see box 2.1), and $L(r_0)$ is the probability calculated for $r = 0.5$ (Morton, 1955). In other words, the numerator expresses the probability of obtaining the results observed (i.e., the probability of the distribution observed of n_j) under the hypothesis of a linkage with an estimated recombination rate r, and the denominator expresses this probability under the hypothesis of independence. The transformation into decimal logarithm helps us to avoid handling very large numbers, and allows immediate interpretation. An LOD of 2 signifies that the linkage is 100 times more probable than independence, an LOD of 3 signifies that it is 1000 times more probable, etc.

What is the threshold that should be chosen for the LOD? When a map is constructed with n markers, $n(n-1)/2$ values of LOD are calculated. Thus, with 100 markers, 4950 tests will be done, and with 300 markers 44,850 tests will be done. If too high an individual type I error is chosen, for example the classical 5%, a high number of linkages will be found significant just by chance. Therefore, an LOD threshold of 3 is usually chosen, which would correspond to a type I error of 0.1% at most. But this value must be modulated according to the number of markers.

These plants are diploids because they are subject to an induced or spontaneous chromosome doubling, but, genetically, each one corresponds to a product of meiosis. Genetic maps can be constructed from populations of such individuals (Zivy et al., 1992; Murigneux et al., 1993).

MEGAGAMETOPHYTE OF GYMNOSPERMS

Unlike angiosperms, gymnosperms have relatively well-developed female gametophytes, the megagametophytes, which occupy the major part of the seed. This enables the extraction of DNA or proteins in a sufficient quantity for multiple analyses. The cells of a gametophyte are genetically identical to each other, since they result from mitosis of a megaspore, which in turn results from meiosis. The megagametophytes of a given individual thus constitute a mapping population that, formally, corresponds to the haploid case. Maps are currently constructed with such material using protein markers (Bahrman and Damerval, 1989; Gerber et al., 1993; Plomion et al., 1995a), RAPD (Plomion et al., 1995b), AFLP (Remington et al., 1999; Costa et al., 2000), and microsatellites (Paglia, 1998) (see also http://www.pierroton. inra.fr/genetics/labo/mapreview.html). This opportunity of performing genetic studies on haploid material also exists in more

---

Box 2.4 Software for genetic mapping

There are several types of software for construction of genetic maps. They differ essentially in the algorithms used for arranging the markers and in the type of progeny considered. Construction of genetic maps from small, non-related families (as is the case in man or breeding animals), for which information from two or three successive generations may be used, relies on software such as Linkage or Crimap. In the case of large progeny resulting from crossing of two inbred lines, as is common in plant genetics, the software most often used is Mapmaker (Lander et al., 1987). Mapmaker takes into account populations of backcross or doubled haploids (a single meiosis analysed), F_2 populations (with dominant or codominant markers), and populations of recombinant lines.

The mapping is sequential. In the first step, Mapmaker computes the recombination rate and the associated LOD scores for all the pairs of markers taken two by two. Fixing an upper limit to the recombination rate and a lower limit to the LOD score, linkage groups are established in which the markers are serially linked. The next step consists of specifying the order of loci within each linkage group. It is impossible to compute the likelihood of all the possible orders as soon as the number of markers in a group exceeds 6 or 7. Therefore, we proceed in steps, establishing the order of 6 or 7 markers that are well distributed in the group, then placing the other markers relative to those already organized. When the order is established, the distances between markers are calculated (in Haldane or Kosambi distance). Several commands make it possible to verify the quality of the map, either through local modifications ("ripple" command), or by analysing the modifications that result from the suppression of each marker of a group ("drop marker" command), or by looking for nearby double crossovers ("error detection").

The robustness of a map will of course depend on the size of the progeny, but also on the presence of markers presenting significant segregation biases, or genotyping errors. In the latter case the distances are artificially increased. Version 3.0 of Mapmaker has a search and correction function for typing errors, while GMENDEL 3.0 (Liu and Knapp, 1990) takes into account distortions of segregation in the F_2.

Finally, Joinmap (Stam, 1993) and GMENDEL software can integrate information from several maps and construct a composite map from several populations. They also make it possible to analyse populations from crosses between heterozygous genotypes.

---

primitive plants, in which the gametophytes are generally even more developed (Pteridophytes, Bryophytes, and algae).

With megagametophytes and doubled haploids, the fact that the markers are dominant or codominant makes no difference, since there are no heterozygous individuals.

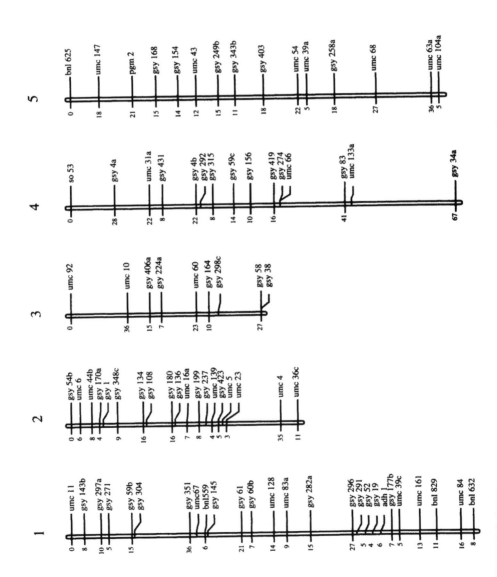

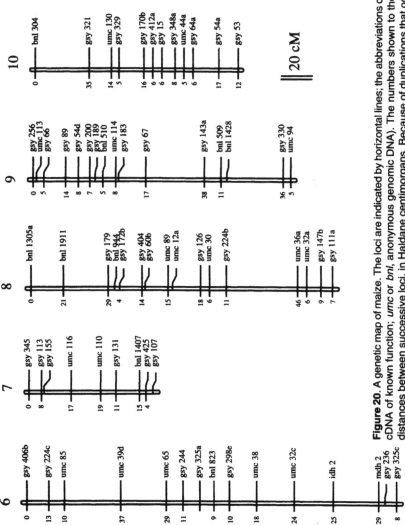

Figure 20. A genetic map of maize. The loci are indicated by horizontal lines; the abbreviations designate the types of probes used (*gsy*, cDNA of known function; *umc* or *bnl*, anonymous genomic DNA). The numbers shown to the left of the chromosomes indicate the distances between successive loci, in Haldane centimorgans. Because of duplications that occurred during maize evolution, certain probes reveal more than one locus. In total, from 134 probes, 150 loci were mapped. One linkage group corresponds to each of 10 chromosomes of this species. The total genetic length of this saturated map is 1815 cM (Causse et al., 1996).

BACKCROSS POPULATIONS

A backcross progeny (BC) results from crossing of an F_1 hybrid with one of its parents (called the "recurrent" parent when a series of backcrosses is being done). If the recurrent parent is genetically fixed, its meiosis does not lead to segregation, and its gametes are genetically identical. On the other hand, the F_1 hybrid will produce gametes of four types already mentioned, which will be the only ones responsible for the segregation observed in the BC progeny (Fig. 21). We can count the proportion of recombinant individuals to estimate r as before, and the distance can be derived. However, it must be noted that there are heterozygotes in this population. Although this does not matter when there is codominance at both marker loci, it does matter if the inheritance is dominant for one of the loci, or both. If there is dominance for one locus, the parental types and recombinant types can be distinguished only if the dominant allele comes from the hybrid (otherwise, the allele of the recurrent parent, present in all the individuals, will mask that of the hybrid parent). If there is dominance for the two loci, the alleles of the two loci must be in a "coupling" phase (i.e., in *cis*, *11/00*, not in "repulsion" phase, or *trans*, *10/01*), *and* the recurrent parent must carry the recessive alleles. There is here a significant limitation of

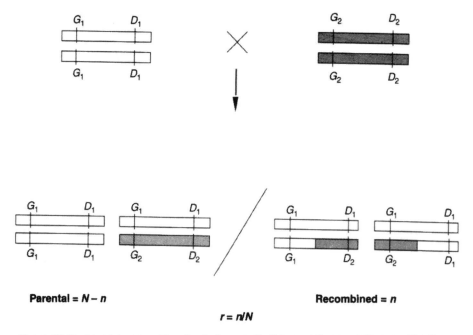

Parental = $N - n$ Recombined = n

$$r = n/N$$

Figure 21. Backcross progeny. To estimate the genetic distances, the populations resulting from backcross correspond *formally* to haploid populations.

dominant markers. The presence of heterozygotes has another disadvantage with respect to QTL detection (see p. 109).

F₂ populations

An F_2 population results from the self-pollination of F_1 individuals (Fig. 22). Compared to the preceding case, two "effective" meioses instead of only one cause the segregation: the male *and* female gametes are subject to recombination. All the possible genotypes can be deduced from the table of gametes (Table 6), in which symmetrical cases must be pooled (the male or female origin of an allele is not known). Nine genotypes are obtained, of which the theoretical frequencies can easily be calculated from the marginal frequencies, which are functions of r (Table 7). It is seen that from the

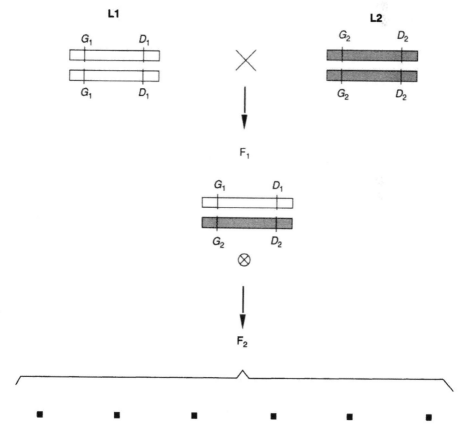

Figure 22. F_2 progeny. The individuals analysed (■) are the result of self-pollination (⊗) of F_1 individuals, or of full-sib crossings. For one locus, there are three possible genotypes, the frequencies of which are 1/4 ($G_1 G_1$), 1/2 ($G_1 G_2$), and 1/4 ($G_2 G_2$). If two loci are considered, there are 9 (3 × 3) possible genotypes, the expected frequencies of which are equal to the product of genotypic frequencies at each locus only if the loci are genetically independent.

Table 6. Table of gametes in an F_2 population

		$\dfrac{1-r}{2}$	$\dfrac{1-r}{2}$	$\dfrac{r}{2}$	$\dfrac{r}{2}$
		G_1D_1	G_2D_2	G_1D_2	G_2D_1
$\dfrac{1-r}{2}$	G_1D_1	$\begin{bmatrix} G_1\,D_1 \\ G_1\,D_1 \end{bmatrix}$	$\begin{bmatrix} G_2\,D_2 \\ G_1\,D_1 \end{bmatrix}$	$\begin{bmatrix} G_1\,D_2 \\ G_1\,D_1 \end{bmatrix}$	$\begin{bmatrix} G_2\,D_1 \\ G_1\,D_1 \end{bmatrix}$
$\dfrac{1-r}{2}$	G_2D_2	$\begin{bmatrix} G_1\,D_1 \\ G_2\,D_2 \end{bmatrix}$	$\begin{bmatrix} G_2\,D_2 \\ G_2\,D_2 \end{bmatrix}$	$\begin{bmatrix} G_1\,D_2 \\ G_2\,D_2 \end{bmatrix}$	$\begin{bmatrix} G_2\,D_1 \\ G_2\,D_2 \end{bmatrix}$
$\dfrac{r}{2}$	G_1D_2	$\begin{bmatrix} G_1\,D_1 \\ G_1\,D_2 \end{bmatrix}$	$\begin{bmatrix} G_2\,D_2 \\ G_1\,D_2 \end{bmatrix}$	$\begin{bmatrix} G_1\,D_2 \\ G_1\,D_2 \end{bmatrix}$	$\begin{bmatrix} G_2\,D_1 \\ G_1\,D_2 \end{bmatrix}$
$\dfrac{r}{2}$	G_2D_1	$\begin{bmatrix} G_1\,D_1 \\ G_2\,D_2 \end{bmatrix}$	$\begin{bmatrix} G_2\,D_2 \\ G_2\,D_1 \end{bmatrix}$	$\begin{bmatrix} G_1\,D_2 \\ G_2\,D_1 \end{bmatrix}$	$\begin{bmatrix} G_2\,D_1 \\ G_2\,D_1 \end{bmatrix}$

Table 7. Expected and observed frequencies (n/N) of different genotypes of an F_2 population (calculated from Table 6)

Homozygotes

$$\begin{bmatrix} G_1\,D_1 \\ G_1\,D_1 \end{bmatrix} = \left(\frac{1-r}{2}\right)^2 = n_1/N$$

$$\begin{bmatrix} G_2\,D_2 \\ G_2\,D_2 \end{bmatrix} = \left(\frac{1-r}{2}\right)^2 = n_2/N$$

$$\begin{bmatrix} G_1\,D_2 \\ G_1\,D_2 \end{bmatrix} = \left(\frac{r}{2}\right)^2 = n_3/N$$

$$\begin{bmatrix} G_2\,D_1 \\ G_2\,D_1 \end{bmatrix} = \left(\frac{r}{2}\right)^2 = n_4/N$$

Heterozygotes

$$\begin{bmatrix} G_1\,D_1 \\ G_2\,D_2 \end{bmatrix} = \frac{r^2}{2} + \frac{(1-r)^2}{2} = n_5/N$$

$$\begin{bmatrix} G_1\,D_1 \\ G_1\,D_2 \end{bmatrix} = r\left(\frac{1-r}{2}\right) = n_6/N$$

$$\begin{bmatrix} G_2\,D_2 \\ G_1\,D_2 \end{bmatrix} = r\left(\frac{1-r}{2}\right) = n_7/N$$

$$\begin{bmatrix} G_1\,D_1 \\ G_2\,D_1 \end{bmatrix} = r\left(\frac{1-r}{2}\right) = n_8/N$$

$$\begin{bmatrix} G_2\,D_2 \\ G_2\,D_1 \end{bmatrix} = r\left(\frac{1-r}{2}\right) = n_9/N$$

observed frequency of each genotypic class, a quadratic equation can be used to calculate an r estimate. The problem posed is thus that of the best estimate of r, which must be resolved by a calculation of maximum likelihood (see box 2.1). The r value being found, it is then, as before, converted into centimorgans. If there is dominance for one of the two loci (6 possible genotypic classes), or dominance for the two loci (4 possible genotypic classes), other likelihood equations will be used, but the principle is the same.

To this type of population must be added the case of "bulk F_3", where *bulked* DNA of individuals resulting from self-pollination of each F_2 individual is analysed. The genotype of each F_2 plant is thus established. If a single plant is analysed per F_3 family, it is also possible to construct a map. For one locus, the proportion is no longer 1:2:1 but 3:2:3, since a heterozygous locus in F_2 has one chance in two of being fixed in the F_3.

Recombinant inbred lines

Populations of recombinant inbred lines (RILs) are obtained by successive self-pollination from F_2 individuals (Fig. 23). At each generation, a single individual is chosen, which will be the parent of the following generation (SSD, single seed descent). This recurrent process has three important genetic consequences.

- As at each generation of self-pollination, each heterozygous locus has one chance in two of being fixed, the mean heterozygosity of the genome decreases rapidly, at a rate of $1/2^n$, n being the number of generations after the F_1. In F_5, 93.75% of the genome is fixed, 96.88% in F_6, 98.44% in F_7, and 99.80% in F_{10}.
- The individuals of a family derived from a given F_2 individual after n generations are all genetically identical, except for the fraction of the genome still heterozygous at the generation $n - 1$.
- Due to the successive meioses, recombination has occurred, and each family derived from a given F_2 individual has fixed a *particular* allelic combination at different loci (hence the term *recombinant* lines) (Fig. 24).

For a given pair of loci, the proportion R of lines that have recombined can be calculated (Fig. 25a). But this proportion *is not* an estimate of r. Indeed, with successive generations, there are several chances for recombination, so that R is higher than r. In 1931, Haldane and Waddington demonstrated that $R = 2r/(1 + 2r)$, whence $r = R/2(1 - R)$ (Fig. 25b).

When self-pollination is impossible, recombinant lines can be constructed from Full-sib crosses. The rate of fixation is simply lower.

Populations derived from non-fixed parents

The populations described earlier are commonly used for mapping in many

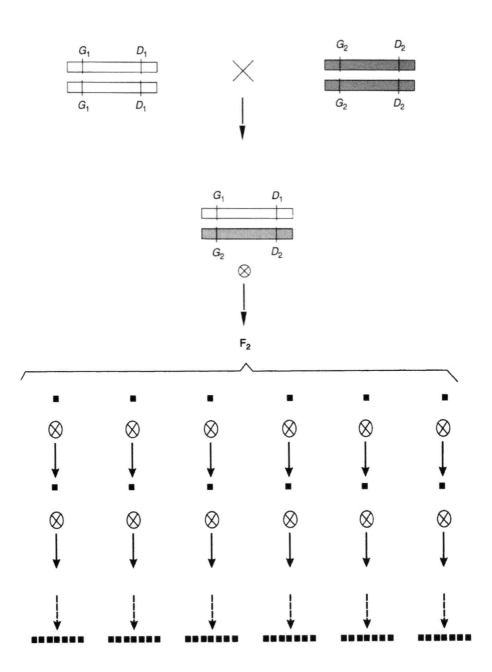

Figure 23. Scheme of generation of a population of recombinant inbred lines. From each F_2 individual, successive self-pollinations (⊗) are realized, choosing only one individual (■) per generation.

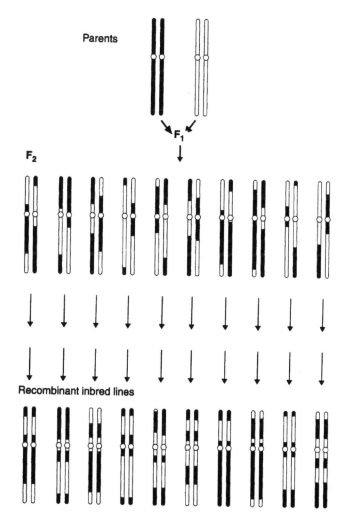

Figure 24. Recombinant inbred lines. Each pair of chromosomes represents the genome structure of a family of recombinant inbred lines derived from a given F_2 individual. The black and white portions represent the contributions of the parental lines. Despite the succession of generations, the genome is not broken into numerous fragments, because at each meiosis the number of crossovers is limited (after Burr and Burr, 1991).

species. However two important particular cases do not come under this approach:

- Species that are self-incompatible, or that are subject to large inbreeding depression and for which pure lines cannot be obtained (e.g., diploid clones of potato).
- Perennial species with a long life cycle, such as trees, in which the

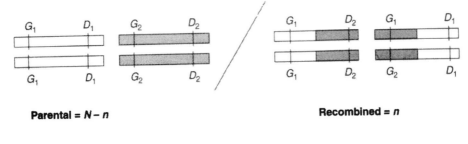

Parental = N − n   **Recombined = n**

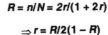

a

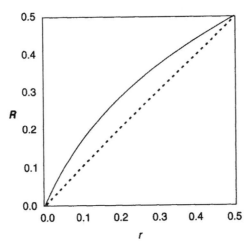

b

Figure 25. Estimate of recombination rates for recombinant inbred lines. a: parental and recombinant genotypes for the loci *G* and *D*, and relationship between the proportion of recombinant lines, *R*, and the recombination rate between these loci, *r*. b: graph of this relationship.

development of mapping populations can take a long time, and where the choice of the genetic material is usually limited (that is the reason why the particular case of gymnosperms, where megagametophytes can be analysed, is so interesting; see subsection *Megagametophyte of gymnosperms*, above).

GENERAL CASE

An F_1 population whose parents are not fixed is peculiar in that it is segregating and may have up to 4 alleles per locus. If two loci are considered, taking into account the different scenarios (coupling vs repulsion, dominance vs codominance, 2, 3, or 4 alleles per locus, and null alleles), there are 27 possible elementary parental configurations (Ritter et al., 1990). For example, $GD/00 \times 00/00$ and $G0/0D \times 00/00$ are the two configurations in which one parent is a double heterozygote (with coupling or repulsion) and the other a recessive double homozygote. $G_1D_1/G_2D_2 \times G_1D_1/G_3D_3$ and $G_1D_2/G_2D_1 \times G_1D_3/G_3D_1$ are the configurations in which there is coupling or repulsion in the two parents for two triallelic loci, etc. As described previously, r can be estimated for all situations using the maximum likelihood method, and then the loci can be arranged (Ritter et al., 1990). It must nevertheless be noted that an ambiguity may exist when both parents are double heterozygotes for dominant loci. Indeed, in this case there may be coupling for the two parents ($GD/00 \times GD/00$), repulsion for both parents ($G0/0D \times G0/0D$), or repulsion for one parent and coupling for the other ($G0/0D \times GD/00$). The first two situations cannot be distinguished from the distribution of phenotypes of the population. Thus, a distribution in which $[GD] = 0.51$, $[G0] = 0.24$, $[0D] = 0.24$, and $[00] = 0.01$ gives $r = 0.2$ if there is repulsion in the two parents, and $r = 0.04$ if there is coupling for one parent and repulsion for the other. When several populations are worked on together, this ambiguity could be removed if other phases are present for these loci. Similarly, if there are three generations, the genotype of grandparents will sometimes eliminate the ambiguity (each time their phase is not ambiguous). But in the rather frequent case where only a single population is available and most of the markers are dominant (as are for example RAPD or AFLP™ markers), all ambiguities may not be removed. Ritter et al. (1990) propose to overcome the problem in two steps:

- Construction of linkage subgroups that contain only non-ambiguous configurations. The most informative configurations are $GD/00 \times 00/00$, $G0/0D \times 00/00$, and $GD/00 \times GD/00$. Two-point and multi-point tests are done *within* each subgroup.
- Union of the subgroups using *codominant* heterozygous loci common to the two parents. The relative orientation of the subgroups is deduced from recombination rates of successive loci. It will thus be noted that even if one uses a technique providing dominant markers, a loose network of codominant markers is very useful.

There are examples of maps constructed from populations generated from non-fixed parents. In diploid potato, for example, Gebhardt et al. (1989) used a population of 67 backcross individuals in which the parental clones were highly heterozygous to construct a map of 680 cM, comprising 141 loci distributed over the 12 chromosomes of this species.

THE DOUBLE PSEUDO-TESTCROSS

In tree genetics, where the mapping populations usually are generated from non-fixed parents, and where dominant markers are often used, a strategy frequently applied is the "double pseudo-testcross" (Grattapaglia and Sederoff, 1994). It consists of constructing *two* maps from the same cross, considering on the one hand configurations of the *GD/00 × 00/00* type, and on the other hand those of the *00/00 × GD/00* type (female × male). In the first case, only segregations due to meiosis of the female parent are observed, so that the genetic map obtained is based only on the recombination rates of this individual. Symmetrically, the other configuration allows construction of the genetic map of the male parent. When this method is used, the difficulties mentioned above are avoided and incidentally the recombination rates between the sexes can be compared. It has been used, among others, in eucalyptus (Grattapaglia and Sederoff, 1994), pine (Kubisiak et al., 1995), and fruit trees (Weeden et al., 1994).

Comparing the various types of populations

The populations described are all currently used for mapping in the plant kingdom (Appendix 1) However, since they have specific genotypic structures and are obtained in specific ways, they differ in several aspects that may have significant practical implications.

PERENNIALITY

Table 8 compares the one-locus genotypes present in the various populations. It can be seen that, unless vegetative multiplication is possible, only progeny of doubled haploids and recombinant lines can be reproduced identically by self-pollination. These "immortalized" populations constitute a reference genetic material on which as many molecular markers can be accumulated as desired, without limitation on the quantity of DNA (or proteins if needed). And of course there is no longer any limitation on the number of traits that can be measured in the successive generations. However, it is worth noting that F_2 or backcross genotypes can be immortalized by self-pollinating, then intercrossing a sufficient number of individuals *within* each family resulting from self-pollination. In genetic terms, each population thus obtained is representative of the F_2 or backcross individual from which it is derived. From analysis of bulks of individuals of each population the genotype of the initial plants can be determined (Gardiner et al., 1993).

Table 8. Genotypes at a diallelic locus (alleles *1* and *2*) in some common mapping populations

Populations	Genotypes
Backcross	*11* (or *22*) and *12*
Doubled haploids	*11* and *22*
Gametophytes	*1* and *2*
F_2	*11, 12* and *22*
Recombinant inbred lines	*11* and *22*

ESTIMATION OF DOMINANCE

Only the F_2 populations, which have the heterozygous genotype together with the two homozygous ones, allow us to estimate the degree of dominance for loci linked to markers (with of course *codominant* markers, see Chapter 4). The backcrosses cannot be used except if populations derived from reciprocal crosses are available, since in that case the three genotypes are represented (again if and only if the markers are codominant).

ACCURACY

For a given size of the progeny, the accuracy of estimation of r differs according to the type of progeny used. The variance of r can be calculated from the likelihood equation (see box 2.1). As shown by Fisher (1937), it is equal to the opposite of the inverse of the second derivative of the likelihood equation:

$$\text{Var}(r) = -dr^2/d^2(\ln L)$$

and is classically written in the form:

$$\text{Var}(r) = 1/Ni_r$$

where N is the number of individuals of the population, and i_r the "individual information", i.e., the gain in precision contributed by each individual of the population. The value i_r depends on the type of population (see likelihood equation, box 2.1) as well as the inheritance at loci (dominance vs codominance). This value, which has been calculated for the most common populations (Allard, 1956), allows comparison of variances of r from various populations (Table 9). The curves in Fig. 26 show that:

- The F_2 with codominant loci provide the most accurate estimations of r, for all the values of r.
- The r values estimated from doubled haploids and backcross have a variance about twice those from F_2 for r values less than 0.15. Intuitively we can say that this is because there was only one effective meiosis at the origin of these populations, and not two as for the F_2. To obtain a given precision, twice as many individuals are thus necessary in backcross or double haploids as for F_2, at least when the map is quite dense. A simple calculation shows, moreover, that when r is very small, the proportion of F_2 individuals having received a recombined gamete from one parent is about $2r$. In other words, there is about twice as much chance of observing a recombination between nearby loci in F_2 as in doubled haploids or backcross. For larger intervals, the values of variances are closer, the possibility of multiple crossovers progressively blurring this difference.
- The recombinant inbred lines approach the precision of F_2 population

Table 9. Individual information (i_r) for various types of populations and types of inheritance at marker loci

Type of progeny	Individual information (i_r)
Recombinant inbred lines	$\dfrac{2}{r(1+2r)^2}$
Backcross, double haploids, gametophytes	$\dfrac{1}{r(1-r)}$
F_2 codominant/codominant	$\dfrac{2(1-3r+3r^2)}{r(1-r)(1-2r+2r^2)}$
F_2 dominant/dominant (coupling)	$\dfrac{2(3-4r+2r^2)}{r(2-r)(3-2r+r^2)}$
F_2 dominant/dominant (repulsion)	$\dfrac{2(1+2r^2)}{(2+r^2)(1-r^2)}$
F_2 codominant/dominant	$2+\dfrac{(1-r)^2}{r(2-r)}+\dfrac{(1-r)^2}{2(1-r+r^2)}+\dfrac{r^2}{(1-r^2)}+\dfrac{(1-2r)^2}{2r(1-r)}$

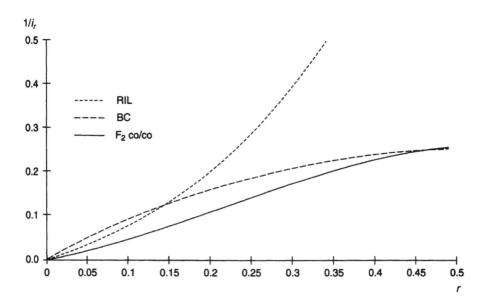

Figure 26. Variance of recombination rate, r. The quantity $1/i_r$ is the inverse of the individual information, which expresses the gain in precision contributed by each individual of the population. Comparison of recombined inbred lines (RIL), backcross (BC), and F_2 when there is codominance at two loci (F_2 co/co).

only for small values of r (lower than ~ 0.05), as if there had been two effective meioses. It is moreover noted here that the relation becomes $R \approx 2r$ for very small values of r, where the same advantage is found as for the F_2. But when r increases, its variance increases quickly, since it goes beyond that of the doubled haploids or backcross towards $r \approx 0.15$, and is three times as high for $r \approx 0.45$. This comes from generations of successive self-pollination, which leads to more possibilities of crossovers, and thus increases the inaccuracy of r.

It is also interesting to compare F_2 populations according to the inheritance at marker loci (Fig. 27). If there is dominance for one of two markers, the variance of r is around twice that observed when there is codominance for the two markers. If there is dominance for the two loci, the situation differs greatly according to the phase. In coupling, the variance is practically the same as in the preceding case for low values of r and then diverges for the high values. In repulsion, unlike in all other situations, the variance is high when r is small, and for $r = 0$ it reaches a value as high as the one observed when $r = 0.5$ with the recombinant inbred lines. This is because the distributions of genotypic frequencies obtained in case of independence or total linkage are very similar (Table 10). Mapping in F_2

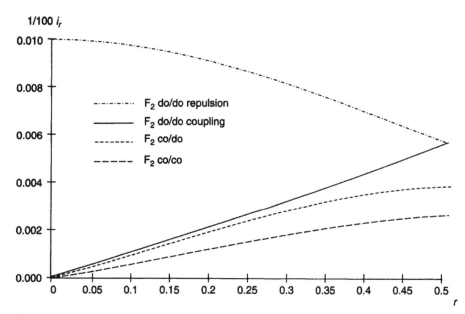

Figure 27. Variance of recombination rate in F_2. The quantity $1/100 i_r$ is the variance of r in a population of 100 individuals. Comparison of populations when there is codominance at both loci (co/co), codominance at one locus and dominance at the other (co/do), dominance at both loci with coupling (do/do coupling), and dominance at both loci with repulsion (do/do repulsion).

Table 10. Segregation in an F_2 population when both loci are dominant. For each locus, allele *1* is dominant over allele *0*. The *phenotypes* are indicated in brackets. In repulsion phase, the estimation of the linkage is imprecise, because the distributions expected in case of independence and total linkage are very close

Population	Coupling	Phase	Repulsion
F_1	*11/00*		*10/01*
F_2 (independence)		9 [11] : 3 [10] : 3 [01] : 1 [00]	
F_2 (total linkage)	3 [11] : 1 [00]		2 [11] : 1 [10] : 1 [01]

with dominant markers is thus clearly less precise than with codominant markers, all other things being equal.

In the case of populations derived from non-fixed parents, the information has also been calculated for the different possible configurations (Ritter et al., 1990). In terms of precision, numerous situations are equivalent to the cases mentioned earlier. To those are added configurations of the $GD/00 \times G0/00$ and $G0/0D \times G0/00$ types, which are close to the case of dominant markers in repulsion in F_2. Configurations of the $G_1D_1/G_2D_2 \times G_1D_1/G_3D_3$ type, which are exactly twice as precise as with the backcross for all values of r, are therefore even more precise than the F_2 with codominant markers for high values of r.

FUNDAMENTALS OF GENETIC MAPS

Saturation of maps

As we have said, a saturated map is one on which every point on the genome is *genetically* linked to at least one marker. How can this objective be attained? When a map is not saturated, there are generally many small linkage groups and unlinked markers. Increasing the number of markers normally results in linking of independent groups, till there are as many linkage groups as chromosomes, and no marker is unlinked. Because the markers are randomly chosen, it is sometimes necessary to test a large number of markers to fill in all the gaps in the map. Moreover, some regions of the genome may be poor in markers because they have high recombination rates. For these recalcitrant regions, rather than accumulating a very high number of markers at random, bulked segregant analysis (BSA) with markers flanking the region may be recommended (Chapter 3 section *Bulked segregant analysis*).

But finding as many linkage groups as chromosomes does not guarantee saturation, which may be demonstrated in two ways. An indirect method consists in accumulating markers. When the genetic length of the map no longer increases, it may be thought that saturation is attained. One difficulty arises here from the fact that, as the number of markers increases, so does the number of genotyping errors, and therefore the estimated length. Control of errors must thus be one of the priorities in genetic mapping (software such as Mapmaker 3.0 has a function to search for errors, based on the detection of recombination events that are "too" close). A more direct method consists in using probes or primers specific to subtelomeric regions (telomeres are made up of repeated sequences located at the ends of chromosomes). If such markers are genetically linked to most distal markers of the chromosomes, then the map is saturated. This task has been accomplished for some chromosomes in maize (Burr et al., 1992). If such probes are not available, BSA can help in more quickly finding markers beyond the most distal markers. Finally, before a map can be saturated, theoretical methods can be used to get an idea of the total genetic length from the number of markers placed on the map and from the type of population (Hulbert et al., 1988; Chakravarti et al., 1991). An illustration of this is found in Gerber and Rodolphe (1994b).

Table 11 summarizes the genetic lengths of genomes of important species for which supposedly well-saturated maps have been constructed. These values range from 630 cM for *Arabidopsis* to 3500 cM for wheat. In relation to haploid numbers of chromosomes, they indicate that the mean genetic length of a chromosome is roughly 100 cM (1 morgan). This signifies that at each meiosis there is *on average* one crossover per chromosome, with large fluctuations according to the chromosomes, and according to the species. Thus, meiosis must be represented as a process that breaks the genome into large fragments. Figure 24 illustrates by example that even in recombinant inbred lines the "shuffling" of the two parental genomes is crude, and the integrity of long portions is preserved.

From the values in Table 11, an order of magnitude of the number of markers to be mapped can be calculated as a function of the degree of saturation desired. With a genome of 1000 cM, 50 markers spaced at 20 cM

Table 11. Relationships between physical and genetic distances in species for which saturated genetic linkage maps are available

Species	Total length (no. of bases* × 10⁹)	Total genetic length (cM)	Length of DNA/cM (kb)	No. of chromosomes
Arabidopsis	0.15	630	140	5
Rice	0.43	1575	280	12
Soya	1.2	2700	440	20
Tomato	0.95	1267	750	12
Bean	0.65	830	780	11
Rapeseed	1.2	1016	1200	19
Maize	2.5	1860	1400	10
Wheat	16	3500	4600	21
Pine	24	1800	13000	12

*Arumuganathan and Earle (1991).

are sufficient in principle. In practice, however, it is necessary to map more markers, because they are not regularly spaced. The reference map (or core map) of maize, published in 1993, comprised only 97 markers, at a maximum distance of 30 cM, which covered nearly all of the 1860 cM of the maize genome (Gardiner et al., 1993). Of course many more markers are now mapped on the maize genome and are organized into "bins" (a bin is an interval between two fixed core marker loci; see http://www.agron. missouri.edu/UMCCoreMarkers.html).

Relativity of map lengths

Genetic distances are not absolute. Various factors have been known to change the recombination rates.

First of all, sex affects the recombination rates, although there is no general rule. In comparing maps resulting from reciprocal backcrosses in tomato, de Vicente and Tanksley (1991) showed that in tomato recombination is larger in females (the map is 20% longer). The same tendency exists in barley (Graner et al., 1991), but in maize (Robertson, 1984) and *Arabidopsis* (Vizir and Korol, 1990) the opposite is observed. There too, the differences between the sexes are highly variable locally.

Genetic distance between the parents also plays a role. Old studies indicate that populations resulting from interspecific crosses lead to genetic distances that are smaller than those obtained with intraspecific crosses (see, for example, Rick, 1969). Maps of molecular markers have amply confirmed this observation. In potato, the map constructed from an intraspecific cross has a size increased by 65% compared to that resulting from an interspecific cross (Bonierbale et al., 1988; Gebhardt et al., 1991). Similar observations were reported in tomato (Paterson et al., 1988) and in rice (Causse et al., 1994) (Fig. 28). The explanation is that the reduction of identity between homologous DNA strands in interspecific crosses leads to a reduction in the frequency of crossovers (Borts and Haber, 1987).

Finally, there is variation in the recombination rates, which are genetically controlled. Along with the possible effects of environment, such variation is undoubtedly the major reason why different populations of the same species do not necessarily display identical map lengths, the variation going beyond 20% (Burr et al., 1988; Beavis and Grant, 1991).

Relations between genetic and physical distances

The low variation in lengths of genetic maps between species contrasts with the considerable variations in quantity of DNA per cell. In the Angiosperms, that quantity varies by a factor of 600 between the extremes, at least for the species studied for this criterion. *Arabidopsis* has only 0.3 pg of DNA per diploid cell (or per 2C value), while a fritillary (Liliaceae) has 255 pg/2C (Bennett and Smith, 1991). The most common 2C values range

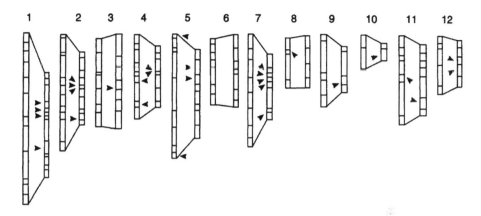

Figure 28. Comparison of genetic maps of rice obtained from intra- and interspecific crosses. To the right, map constructed from an interspecific backcross between *Oryza sativa* and *O. longistaminata*. To the left, map constructed from an F₂ within *O. sativa*. The arrowheads indicate the intervals between markers from markers that are significantly different between the two maps. The intraspecific map is overall 25% longer than the interspecific map (after Causse et al., 1994).

between 1 and 10 pg. The Gymnosperms have fairly high average values, between 10 and 60 pg/2C. As the number of genes expressed must be of the same order of magnitude within the higher plants, these differences in quantity of DNA would be essentially due to highly significant variation in the quantity of non-transcribed repeat DNA. Beyond the possible adaptive significance of this variation, which is still debated, the important point is that it is not reflected in the total genetic length of genomes. In other words, it is clear that the frequency of crossovers per unit of physical length decreases when the genome size increases. This can also be verified between chromosomes of a single species: in yeast, the smallest chromosome has on average twice as many recombinations per kilobase as the largest (Kaback et al., 1992). These results are consistent with observations that the regions of heterochromatin, containing highly repeated sequences, correspond to zones of low recombination (Flavell, 1985). As a corollary, studies on yeast have indicated a correspondence between the recombination and zones of transcription (Oliver et al., 1992).

The average number of base pairs per centimorgan will thus depend on the species under consideration. Table 11 shows that, from *Arabidopsis* to pine, the values vary by a factor of almost 100, from 140 kb to 13,000 kb. This information is important for positional cloning. To find a marker at 1 cM of a gene of interest may be sufficient to start off the chromosome walking or landing in a species with a small genome (Chapter 3, section *Principle of positional cloning*), but it cannot be used in maize or wheat. This conclusion may be qualified, however, because the values are averages over the entire genome, and very large local variations of recombination rates may exist. Dooner (1986) observed, for example, that one centimorgan in the neighbourhood of the *bronze* locus in maize corresponds to 14 kb, a value 100 times as small as the mean value for the entire genome. Inversely, in the neighbourhood of the centromeric region, where there is a restriction of recombination, the number of kilobases per centimorgan can be much greater than on the rest of the genome. In *Arabidopsis*, Schmidt et al. (1995) compared the relative distances of physical (YAC) and genetic (RFLP) maps of chromosome 4 and showed that significant differences could locally be observed (Fig. 29).

Analysis of genome organization

In species such as maize, rapeseed, or rye, a significant percentage of probes reveal more than one locus. Their mapping may reveal homologies *within* the genome, in the sense that the order of markers on a linkage group (or one portion of the linkage group) is found on another group (Fig. 30). These homologies may be interpreted by supposing that the genomes in question have undergone partial or total duplication in the course of evolution (Helentjaris et al., 1988; Slocum et al., 1990; Song et al., 1991). This phenomenon may even occur in species with a small genome, such as rice (Kishimoto et al., 1994) and *Arabidopsis* (Blanc et al., 2000; Grant et al., 2000).

Comparative mapping

The idea of comparing the genetic maps of different species is an old one, especially in animals. In 1920, Dunn compared the linkage between the gene for albinism and the pink-eye gene in rat and mouse (cited in Lyon, 1990). Several arrangements of genes that are homologous between the maps of mammals have since been described. Some linkage groups are even conserved among the vertebrates, from fish to man (Morizot, 1990).

The possibility of using heterospecific probes, along with the relative facility of constructing maps in plants, has allowed numerous comparisons of the order of genes in species belonging to a single botanical family. In the Solanaceae, Bonierbale et al. (1988), then Tanksley et al. (1992) showed that the genomes of tomato and potato were colinear, with the exception of 5 paracentric inversions. There is thus a very good conservation of *synteny*. (There is synteny when loci are carried on the same chromosome, whether or not they are genetically linked. *Asynteny* pertains to loci carried by different chromosomes.) Between tomato and pepper, which are phylo-genetically not so close, rearrangements are more numerous (around 20), but locally the order of genes remains conserved (Tanksley et al., 1988; Lefebvre et al., 1995; Livingstone et al., 1999). In the Fabaceae, comparison of maps of lentil and pea reveals 8 well-conserved regions, which represent 40% of the genome (Weeden et al., 1992). But it is the Poaceae that have been

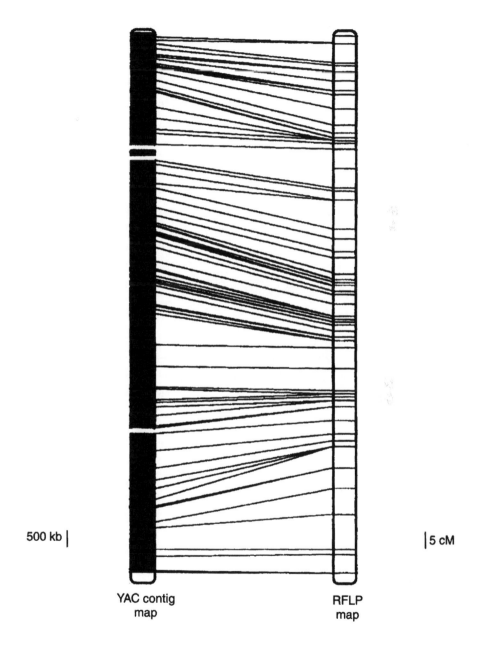

500 kb | |5 cM

YAC contig   RFLP
map   map

Figure 29. Correspondence between the physical map (YAC contig map) and genetic map (RFLP map) of chromosome 4 of *Arabidopsis thaliana*. The 80 RFLP markers were mapped on up to 100 recombinant inbred lines, using the Kosambi mapping function (modified from Schmidt et al., 1995).

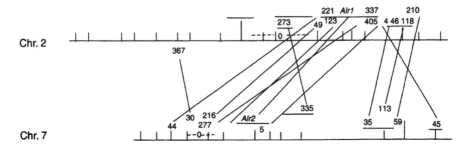

Figure 30. Example of regions duplicated between chromosomes 2 and 7 of maize. Using probes that reveal two loci in place of one, the order of markers is seen to be locally conserved (Helentjaris et al., 1988).

the focus of the most exhaustive studies. In the Triticae tribe, the order of markers is largely conserved between wheat, barley, and rye (Sharp et al., 1988; Devos et al., 1992; Wang et al., 1992), even if the rye genome seems to differ from that of wheat by many translocations. In the Andropogonae tribe, a good conservation of linkage groups and the order of markers has been observed between maize and sorghum (Whitkus et al., 1992; Melake Berhan et al., 1993), and the sugarcane genome could be related to this group, despite its high ploidy level (Grivet al., 1994) (Fig. 31). The linkage groups found in sorghum and sugarcane have generally two homologous regions in maize, which, as already mentioned, has undergone ancestral duplication. However, maize seems to have undergone more rearrangement than sorghum and sugarcane since these three species diverged. Comparisons between tribes within the Poaceae are also possible. Rice, wheat, and maize share several colinear genomic regions (Ahn and Tanksley, 1993; Ahn et al., 1993a; Kurata et al., 1994). A synthesis of data on Poaceae was proposed by Moore et al. (1995) and by Devos and Gale (1997, 2000). From a fundamental point of view, these results, once refined and extended to more species, could be used to analyse the evolution of the genome structure from the ancestral genome of the families of Poaceae.

Even though the probes generally used in experiments are largely anonymous, the idea that the positional homology of genes is associated with functional homology has long been acknowledged. To take classical examples, let us cite the mutations that determine the loss of the ligule, mapped in maize and rice (Ahn et al., 1993a), or a major gene for frost tolerance and needs of vernalization found in rye, wheat, and barley (Plaschke et al., 1993). Comparative mapping can constitute a major tool for integrating research programmes that are working in parallel lines on different species. This point was advocated in particular by Bennetzen and Freeling (1993) and Freeling (2001), who proposed that grasses be considered as "a single genetic system". The following approaches may be possible:

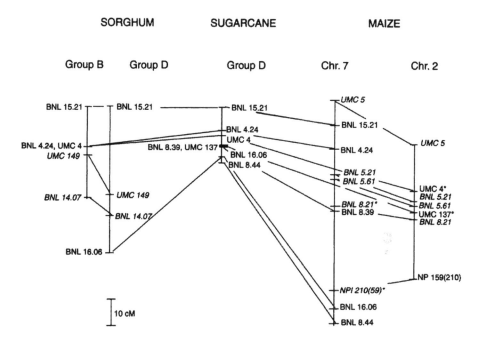

Figure 31. Comparison of the order of loci and recombination rates for various linkage groups in sorghum, sugarcane, and maize. Probes of maize have been used on the other two genomes (Grivet et al., 1994).

- Markers from a species can be used on related species to saturate a given region. This could prove particularly useful in species with a high ploidy level, where mapping the entire genome is not feasible (Asnaghi et al., 2000).
- The gene homologous (orthologous) to the desired gene in a related small-genome model species for which a YAC or BAC library is available can be cloned, and then the sequence obtained can be used to isolate the gene in the species studied (Robert et al., 1998).

3

Mapping of Major Genes

D. de Vienne

APPROACHES TO MAPPING OF MAJOR GENES

When a trait is controlled by a single gene, the markers linked to that gene can be found by construction of genetic linkage maps (see Chapter 2). However, it is not judicious to map the entire genome to look for markers on a very limited region. Besides, the probability of finding markers tightly linked to the gene of interest using such a method is negligible. There are two ways to save considerable time and money on this task: near-isogenic lines and bulked segregant analysis.

Near-isogenic lines

Isogenic lines are lines that have the same genotype. Lines that are near-isogenic for a given locus are lines that have the same genotype with the exception of a locus that has fixed different alleles. This case is rare in practice, since a mutation must occur and must be fixed. To develop similar material, successive backcrosses may be done between a "donor" line that has a desired allele and a recipient (or recurrent) line that has another allele, then the material is fixed by means of self-fertilization. For the chromosome that carries the introgressed allele, this process leads to moderate isogenicity: at each backcross generation, the selection of individuals that have the desired allele results in "hitch-hiking" of a part of the chromosome carrying the gene (Fig. 32).

With such material, markers linked to the gene can be found by comparing the donor line D, the recurrent line R, and the line that has received the gene, R' (R and R' are near-isogenic). For a given marker, three situations are expected:

- D, R, and R' are identical: the locus is monomorphic, not informative (Fig. 33a).
- D is different from R, but R and R' are identical: the allele of D has not been introgressed in R' (Fig. 33b). The marker is thus not linked to the

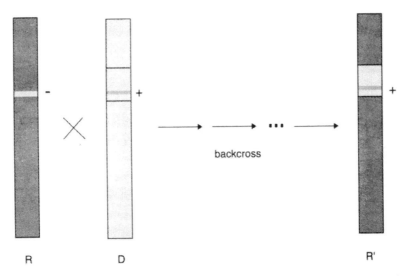

Figure 32. Near-isogenic lines. During the process of successive backcrosses aimed at replacing an unfavourable allele (–) of the "recipient" line (R) by a favourable one (+) of the "donor" line (D), a portion of the genome of D is carried with the introgressed allele.

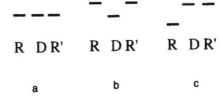

Figure. 33. Possible profiles when comparing near-isogenic lines. R: recipient line. D: donor line. R': line that has received a portion of genome of D. a: non-polymorphic marker. b: polymorphism outside the introgressed region. c: polymorphism between R and R' due to the introgressed region.

introgressed gene.

– R and R' are different, and R' is identical to D (Fig. 33c): the allele of D has been introgressed in R'. The marker is thus linked to the introgressed gene.

The first two situations are expected to be the most frequent, since only a small fraction of the total genome is introduced.

Comparison of R and R' might be sufficient to search markers of the gene under study. But D must be used as a control, to be certain that the allele of R' comes from D, and not from technical (artefactual bands) or genetic causes (gene contamination by cross-pollination).

Any type of marker can be used for these comparisons, but the best techniques are obviously those of genetic fingerprinting, which can provide several tens of markers in a single experiment: AFLP™, ISSR, etc. The fact that these markers are usually dominant is not a problem, since for either phase, coupling or repulsion, the differences can be detected. If required, codominant markers can be derived from dominant markers (see SCAR in Chapter 1). Once the markers have been found by this method, they must be mapped in order to verify that they are really linked to the gene being studied.

Bulked segregant analysis (BSA)

Several major genes, especially disease-resistance genes have been mapped using near-isogenic lines (Diers et al., 1992; Young et al., 1988; Sarfatti et al., 1989; Martin et al., 1991; van der Beek et al., 1992; Paran et al., 1991; Hinze et al., 1991; Yu et al., 1991; Penner et al., 1993; see Michelmore, 1995, for a review). The creation of this material is still time-consuming, because several generations are required, and sometimes "false positives" are detected (a segment of D is introgressed but does not carry the desired gene). For this reason, the technique of bulked segregant analysis (BSA) is used (Arnheim et al., 1985; Michelmore et al., 1991). This technique consists in comparing bulks of DNA of individuals having the same genotype. For example, from an F_2 population of recombinant inbred lines or doubled haploids segregating for a resistance gene, differences between a bulk of individuals *RR* (homozygous resistant) and a bulk of individuals *rr* (homozygous susceptible) are looked for (Fig. 34). Only the markers linked to the gene show differences, either of band position (codominant markers) or of presence/absence (dominant markers; fingerprint methods are again the most convenient for BSA). As described before, these linkages must be verified by mapping the markers identified.

The minimum size of the bulks must be determined by imposing a maximum probability of declaring an unlinked polymorphic marker linked to a gene. This probability depends on both the type of population and the inheritance at marker. For example, for a dominant marker segregating in an F_2 population, the probability of finding an independent marker linked to the gene with bulks of size n is $2(1/4)^n[1 - (1/4)^n]$. With only 4 individuals per bulk, the proportion of false positives is only about 1%, and with 10 individuals it is 2×10^{-6}. Therefore, 10 to 15 individuals are generally used. On the other hand, BSA does not necessarily give all-or-nothing responses. Differences of band intensity between bulks may be observed, which are due to the fact that one or a few recombinant individuals may be present in the bulk as soon as the linkage is not total between the marker and the gene.

BSA has been successfully used to map major genes in plants, for example, a gene for sex determinism in pistachio (Hormaza et al., 1994) and a gene for restoration of fertility in rapeseed (Delourme et al., 1994). More results can be found for disease-resistance genes, for example, in lettuce (Michelmore et al., 1991; Kesseli et al., 1993; Maisonneuve et al., 1994), bean (Haley et al., 1993), or sugar pine (Devey et al., 1995). Additional examples may be found in Michelmore (1995a).

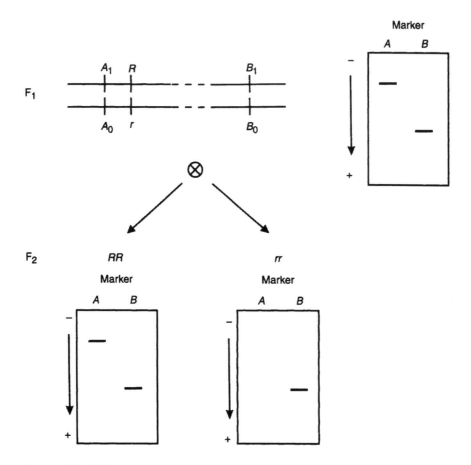

Figure 34. The BSA method. *R* and *r* designate two alleles of a locus to be mapped, *A* designates a marker linked to that locus, and *B* an independent marker. 1 and 0 designate the alleles "presence" and "absence" of band, respectively. In the progeny, individuals with the same genotype are analysed together.

This method can also be used to fill in gaps in the maps. The bulks can be constituted on the basis of genotypes of markers that flank the interval to be filled in, considering only the non-recombinant individuals between these loci (i.e., the G_1D_1/G_1D_1 individuals are compared with the G_2D_2/G_2D_2 individuals). Similarly, markers of the ends of chromosomes can be found by BSA from the most distal markers of the linkage groups. Thus we can reach step by step the point at which markers that "lengthen" the map are no longer found (Reiter et al., 1992).

In principle, BSA and near-isogenic lines are related. Bulks look like near-isogenic lines with a heterozygous genome, because they contain the two parental alleles in statistically equal proportions for the loci not linked to the gene of interest. Nevertheless, as there is usually dominance at the locus studied and at the marker, the markers that are in repulsion with the gene cannot be detected in bulks of backcross and F_2 even if they are totally linked, since a band will appear in the two bulks (Table 12). Thus, on average, twice as many markers must be tested as with doubled haploids or with recombinant inbred lines. One way to overcome the problem is to compare the bulk *rr* having the band to the *parent RR*, which does not show it.

Table 12. Bulked segregant analysis (BSA) for finding markers linked to a major gene. Comparison of progeny when there is dominance at the locus studied *and* at the marker. *R* and *r* represent respectively the dominant and recessive alleles, and *1* and *0* represent the presence and absence of band for a marker that is totally linked to the gene. The phenotypes of the two groups obtained in each situation are indicated within brackets.

| Progeny | Phase | | Informative markers |
	Coupling	Repulsion	
F_1 (genotypes)	*R1/r0*	*R0/r1*	
Doubled haploids	[R1] and [r0]	[R0] and [r1]	All
Backcross[1]	[R1] and [r0]	[R1] and [r1]	Only if coupling
F_2	[R1] and [r0]	[R1][2] and [r1]	Only if coupling
Recombinant inbred lines	[R1] and [r0]	[R0] and [r1]	All

[1]   Backcrossing is done on the parent homozygote for the recessive allele of the gene studied (otherwise no segregation would be observed).
[2]   This phenotypic class is heterogeneous with individuals [R1] and [R0] in proportion [2:1].

USE OF MARKERS FOR CLONING MAJOR GENES

Principle of positional cloning

A gene with unknown messenger RNA or protein may be isolated by transposon insertional mutagenesis or positional cloning. Transposon insertional mutagenesis takes advantage of the capacity of transposable elements to inactivate the genes where they insert. Once a mutant has been found by its phenotypic effect, the gene can be isolated via the transposable element responsible for the mutation. This naturally can be done only in species in which there are genetic materials able to induce transpositions, as in maize.

Positional cloning, which consists of isolating a gene using tightly linked markers, has a more general application. In the initial method, or "chromosome walking" (Rommens, 1989), the desired gene is approached progressively using overlapping large genomic clones (cosmids, YAC, BAC) (Burke et al., 1987; Ward and Jen, 1990). Besides the difficulty of testing each new clone for the presence of the gene, this approach may take considerable time in higher plants because of the size of their genomes. Even if a marker 1 cM distant from the desired gene is available, this may represent several hundreds or thousands of kilobases to be scanned, which is almost impracticable (Table 11). "Chromosome jumping" may be an improvement on this technique (Rommens, 1989). But the techniques for finding markers based on near-isogenic lines or BSA, along with the fine mapping methods (see below and box 4.2 in Chapter 4), provide more powerful tools. The idea is that several thousands of markers can be screened on such material, in order to search for markers at a genetic distance corresponding to the physical size of YAC, BAC, or cosmid clones. In screening libraries of such large fragments with these markers, we can thus hope to pinpoint directly the clones containing the gene of interest, at least in species with a small genome. This approach is sometimes referred to as "chromosome landing" (Tanksley et al., 1995).

High-resolution mapping

Once polymorphic markers are selected, their mapping order must be determined to find the one closest to the gene under study. High-resolution mapping must be performed, which requires the analysis of a great number of individuals, for several tens of markers (if, for example, 8000 markers have been screened in a genome of 2000 cM, around 40 markers are expected in the interval of 10 cM around the target gene). To obtain a resolution less than a tenth of a centimorgan, which is the minimum for finding the clone that possibly carries the gene, more than 3000 backcross individuals are necessary if one wishes to detect at least one recombination with a probability of 0.95. One way to reduce the work is to begin by analysing the individuals of the population for only two markers that are known to flank the gene, then to retain for further analysis with other markers only those individuals that have recombined. Another strategy is based on the analysis of bulks of individuals (Churchill et al., 1993). It consists of separating the individuals of the population according to their genotype at the locus studied, then analysing the DNA of bulks of individuals. From the proportion of bulks with profiles that reveal at least one crossover, the recombination rate between the marker and the target gene can be calculated. As each bulk is analysed for a series of markers, three-point tests can be done, and the markers closest to the gene can be identified.

Looking for the target gene in genomic clones

The markers most tightly linked to the gene are used to screen a large fragment library (YAC, BAC, or cosmid). The positive clones are supposed to contain, among other genes, the target gene. Bioinformatic analysis of the clone may reveal relevant candidates. The clone may be used to screen a cDNA library, which also provides candidates. To decide between the candidates, the most direct method is transformation, whenever possible. Another approach is high-resolution mapping of the clone in order to identify the gene that co-segregates with the trait. Finally, for some traits, studies of expression may help to decide. For example, the expression of a gene involved in a response to stress must be induced, or controlled, by that stress.

Positional cloning has thus been used successfully to isolate genes in various plant species, usually with a small genome: examples are the gene *ABI3*, responsible for reduction of responses to abscissic acid in *Arabidopsis* (Giraudat et al., 1992), a gene coding for an omega-3 desaturase in *Arabidopsis* (Arondel et al., 1992), the gene *Pto* of resistance to *Pseudomonas syringae* in tomato (Martin et al., 1993), the gene *AXR1* for insensitivity to auxin in *Arabidopsis* (Leyser et al., 1993), etc. For disease resistance genes and root symbiosis genes, various reviews are available (Michelmore, 1995b; Hammond-Kosack and Jones, 1997; Jones, 2001; Stougaard, 2001).

4

Mapping and Characterising Quantitative Trait Loci

D. de Vienne and M. Causse

Formal genetics deals with traits displaying discrete (or discontinuous) variation, such as round or wrinkled seeds in pea, white or red colour of petunia flower, and normal or dwarf size. There is an incalculable number of such variations, which are usually controlled by single genes, and in which a "wild" phenotype is distinguished from a "mutant" phenotype. Molecular analyses show that the mutant phenotypes are most often due to major alterations of a gene (for example, the *r* gene of pea, which determines the wrinkled form of the seed, is inactivated by an insertion; Bhattacharyya et al., 1990). Such mutations are actually rare in natural populations, because they are generally strongly counter-selected as a result of their large pheno-typic effects.

On the other hand, many traits display *continuous* variation in populations. For such so-called quantitative traits, the individuals cannot be *classified*, as they are for discrete variation, but are *measured*. This continuous distribution may be explained, without violating the laws of Mendelian genetics, by assuming simply that *several* segregating loci influence the variation of the trait (Fig. 35), to which environmental effects may also contribute (Fig. 36). Since the early 1980s, these loci have usually been called QTLs (quantitative trait loci). Nevertheless, this term should not obscure the fact that the loci involved are not different in nature from those responsible for discrete variations. The working hypothesis is that *moderate differences of effects* between "wild" (or active) alleles are responsible for the variation of quantitative traits, rather than a "mutant-wild" opposition. Studies on QTLs for maize height support this hypothesis, since many of them map in the same regions as dwarf genes (Helentjaris, 1987; Beavis et al., 1991). For instance, in various maize progeny, a QTL for height is found in the region of the dwarf gene *d3*, on chromosome 9. There is probably a single locus, with differences in allelic effects that account for the QTL, and a loss-of-function allele for the mutant

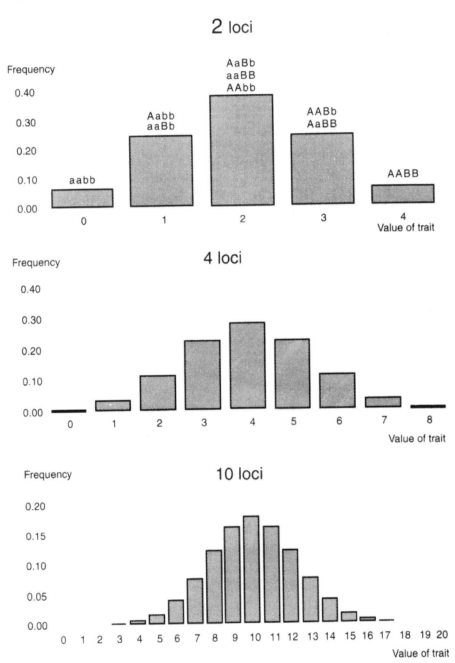

Figure 35. Distribution of genotypic values in an F_2 population according to the number of independent segregating loci. The value of a genotype is supposed to be the sum of contributions of each allele at each locus, one carrying a null value and the other a value of 1. There are no environmental effects.

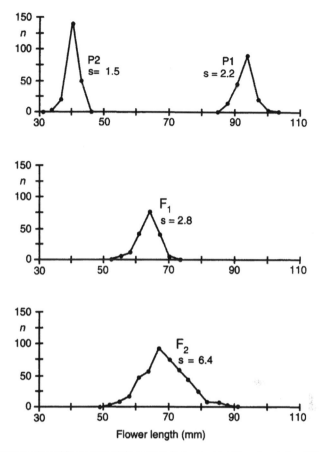

Figure 36. Distribution of flower length in a cross between two inbred varieties (P1 and P2) of an ornamental species of tobacco. s: standard deviation (after East, 1915). The parents, like the F_1's, are genetically homogeneous, therefore their standard deviation can only be due to environmental effects. Flower length (mm).

phenotype (Touzet et al., 1995b). The characterisation of some genes with quantitative effects in plants supports this conception (Byrne et al., 1996; Doebley et al., 1997; Yano et al., 2000).

Till recently, the genetics of quantitative traits, or quantitative genetics, based on statistics, could not provide a general description of genetic properties of these traits except with heavy hypotheses: approximate number of genes, relative weight of additive variances, dominance and epistasis in total variance, and prediction of results of selection from the resemblance between relatives. Even though this science had achieved great success in genetic improvement of plants and animals, it remained unable to estimate the *individual* effects of QTL, their intra-locus (dominance) and inter-locus (epistasis) interactions, and their position in the genome, and it could not provide any means for their molecular identification. The principle of QTL mapping was known long ago. In 1923, Sax mentioned an association between the colour (monogenic trait) and the weight (polygenic trait) of bean seeds and suggested that the gene controlling the colour was genetically linked to one or several factors controlling the weight. Other examples of the same type were subsequently published but have never resulted in a systematic method of finding QTL. As mentioned in Chapter 1, markers with visible effect can only be present in small number in a given genetic background and may interfere with the quantitative trait studied (to take an extreme example, dwarf phenotype is very difficult to use as a marker for detecting a QTL for height). Isozyme markers have given rise to preliminary illustrations of the principles of analysing QTL proposed by Thoday (1961), but it is the molecular markers, which can saturate the genome, that have led to the emergence of a new approach in quantitative genetics.

PRINCIPLE OF QTL MAPPING

Mapping of QTL is based on a systematic search for linkage disequilibrium between marker loci and QTLs. For this, we need the following:

- To have a segregating population (in which there is a relationship between linkage disequilibrium and genetic distance). As for genetic mapping, the most efficient populations are those resulting from crosses between two homozygous lines.
- For each individual of the population, to determine the genotype for a series of marker loci distributed throughout the genome (which allows construction of a saturated genetic map).
- To measure the value of the quantitative trait for each individual of the population.
- To use biometric methods to find the markers whose genotype is correlated with the trait(s), and estimate the genetic parameters of the QTL(s) detected.

There are three ways to assign an estimate of the value of the trait for each individual:

- Take the value of the individual itself (in that case the trait must have high heritability), or the average of genetically identical individuals (which is possible with doubled haploids or recombinant inbred lines).
- Take the mean value of the family derived from the self-fertilization of the individual. The increase in precision thus would allow us to limit the number of individuals to be genotyped. But in the case of F_2 the estimated degree of dominance will be diminished by half (Stuber et al., 1987; see below).
- Take the average of the progeny of a cross between the individual and a homozygous "tester" genotype. This last case is interesting in plant

breeding, when hybrid varieties are to be developed. The dominance cannot be estimated in that case, since there are no more heterozygotes Q_1Q_2, but both genotypes Q_1T and Q_2T (1:1), where T is the allele of the tester.

Biometric techniques to find markers correlated to traits can be classified in two categories, those that are based on marker-by-marker analyses, and those that simultaneously take into account two or more markers.

Detection of QTL considering markers individually

This approach, first developed by Soller et al. (1976) and used since the early 1980s (Tanksley et al., 1982), is based on analysis of variance or Student's t-test. The principle consists simply in looking for a relation between the genotype of *each* marker locus and the value of the quantitative trait considered (see Edwards et al., 1987; Keim et al., 1990; Stuber et al., 1992; Dirlewanger et al., 1994, among others).

Finding QTL by analysis of variance in an F_2 population

The case of F_2 progeny is the most convenient to illustrate the search for QTLs on a marker-by-marker basis, because it is the only one in which the three possible genotypes at one locus are present. The other common populations are mentioned in the section *Comparison of F_2, backcross, doubled haploids, and recombinant inbred lines.*

For a codominant marker locus M, the individuals of an F_2 population are distributed into the three genotypic classes M_1M_1, M_1M_2 and M_2M_2. The average of each class for the quantitative trait studied is then calculated, and an analysis of variance is performed to find whether there are significant differences between these averages (Fig. 37; box 4.1). This is done for each marker of the population (each marker enables a specific partition of the population), so that one can construct a map of genomic regions where there is an association between a marker and the quantitative trait. The simplest way to explain this effect is illustrated in Fig. 38. If the marker is genetically linked to a polymorphic QTL, the partitioning of the progeny from the marker locus corresponds to the partitioning from the QTL (the consequence of the recombination is analysed in box 4.2). A significant difference between the means signifies that the effects of two alleles of the QTL are sufficiently different to have detectable consequences. Hereafter, according to an "economical" interpretation, we will assume that there is only *one* QTL close to the marker. But it may naturally occur that two nearby QTL "mimic" the action of a single one, and only fine mapping studies can separate them (see box 4.3) (Paterson et al., 1990; Eshed and Zamir, 1995).

If the recombination rate between the marker and the QTL is zero, the two parameters that define the QTL action, the additive effect a and the degree of dominance d, are easy to define. The additive effect a is equal to

Box 4.1: Detection of QTL by analysis of variance in an F_2 population

The model used to detect a QTL in an F_2 population is a one-way model of analysis of variance with fixed effects, where the marker locus is the factor with three levels, i.e., the three genotypes M_1M_1, M_1M_2, and M_2M_2:

$$Y_{ij} = \mu + \alpha_i + E_{ij}$$

where Y_{ij} is the phenotypic value of the jth individual of the ith genotype ($i = 1, 2, 3$), μ the mean of the population, α_i the effect of genotype i, and E_{ij} the residual error resulting from segregation of other QTLs and environmental variance.

The factor (marker) sum of squares (SS_M), the error sum of square (SS_E), and the mean squares (MS_M and MS_E) are conventionally calculated:

$$SS_M = \sum_i n_i (y_{i.} - y_{..})^2$$

$$MS_M = \sum_i n_i (y_{i.} - y_{..})^2 / (g-1)$$

$$SS_E = \sum_{ij} (y_{ij} - y_{i.})^2$$

$$MS_E = \sum_{ij} (y_{ij} - y_{i.})^2 / (N-g)$$

where n_i is the number of individuals of genotype i ($j = 1, 2, ..., n_i$), g is the number of genotypes at the marker locus (here $g = 3$), and N is the total number of individuals of the population.

For the test, the value of the ratio $MS_M/MS_E = f$ is compared to the threshold value read in a table of F for $(g-1)$ and $(N-g)$ degrees of freedom and a type I error α. If f is higher than the threshold value, the conclusion will be that there is one (or more) segregating QTL linked to the marker.

The fraction of phenotypic variation explained by the marker (i.e., by the QTL that is linked to it), called R^2, is:

$$R^2 = SS_M/(SS_M + SS_E)$$

For example, an $R^2 = 0.15$ signifies that 15% of the *phenotypic* variation is due to the segregation of the QTL. The *genotypic* R^2 is necessarily higher, due to the environmental variance that increases the denominator (in SS_E, which also comprises the effects of other QTLs). The *genotypic* R^2 can be calculated in cases where replications or offspring of the genotypes are available, which allow the part of variation due to the environment to be estimated. Naturally, the notion of R^2 is valid for any type of population.

The value of f of the analysis of variance is correlated to the effect of the QTL but cannot constitute a measure of this effect, because it depends on the number of individuals analysed. There is the relation $1/R^2 = 1 + (1/f)(N-g)/(g-1)$.

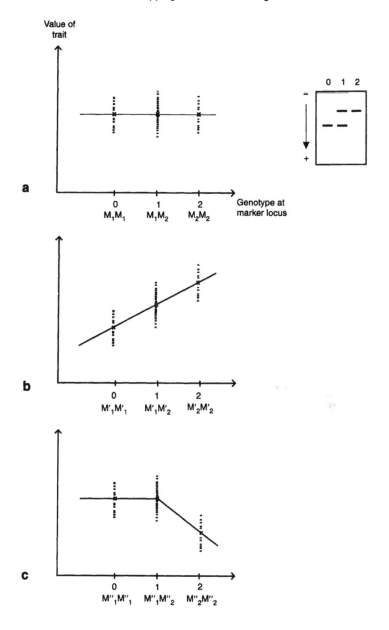

Figure 37. Principle of QTL detection in an F_2 population with a search marker by marker. M, M', and M" are three independent marker loci. The subscripts 1 and 2 designate the alleles. The X-axis indicates the number of doses of the allele M_2 in the three genotypes. a: No difference between means (on the right are shown the electrophoretic profiles of the three genotypes at the marker locus). b: Significant difference between means with additivity. c: Significant difference between means with dominance: The mean of heterozygotes is equal to the mean of the homozygotes $M''_1M''_1$.

Box 4.2: Relation between additive effect for the marker and additive effect for the QTL, and principle of interval mapping

For simplicity, we will take $m = (m_{11} + m_{22})/2$ as origin of the genotypic values of the QTL. Thus we have $m_{11} = m - a$, $m_{12} = m + d$ and $m_{22} = m + a$, where a is the additive effect and d is the degree of dominance of the QTL. The values of three genotypes Q_1Q_1, Q_1Q_2, and Q_2Q_2 will thus be respectively $-a$, d, and $+ a$. If we consider a marker linked to this QTL with a recombination rate r, the genotypic values *at the marker locus* can be calculated according to the following reasoning. The genotypic value of individuals M_1M_1, G_{11}, will be $-a$ in the absence of recombination, i.e., with a probability $(1 - r)^2$, d if there has been recombination for one of the two gametes, i.e., with a probability $2r(1 - r)^2$, and a if the two gametes are recombined, i.e., with a probability r^2. Thus, we get:

$$G_{11} = - a(1 - r)^2 + 2r(1 - r)d + ar^2$$

$$= - a(1 - 2r) + 2r(1 - r)d$$

Using a similar reasoning we obtain:

$$G_{12} = d[1 - 2r(1 - r)]$$

$$G_{22} = a(1 - 2r) + 2r(1 - r)d$$

The line passing through the genotypic values of homozygotes at the marker locus has the slope:

$$a' = (G_{22} - G_{11})/2$$

From this we get

$$a' = a(1 - 2r)$$

The degree of dominance for the marker is

$$d' = G_{12} - (G_{11} + G_{22})/2$$

from which we get

$$d' = d[1 - 4r(1 - r)]$$

and thus

$$d' = d(1 - 2r)^2$$

It can be seen that d decreases more rapidly than a when r increases.

This type of reasoning is applied in one of the approaches used to find the position of a QTL within the interval between two markers, or "interval mapping". The following scheme shows a QTL Q between markers A and B, with r_1 the recombination rate between A and Q, and r_2 between Q and B. The double crossovers are neglected, and therefore $r \approx r_1 + r_2$, where r is the recombination rate between A and B (this approximation is better for small r).

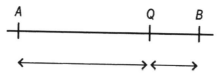

The example of backcross progeny is the simplest to show how to estimate r_1 (or r_2). Consider the cross of hybrid F_1 A_1B_1/A_2B_2 with parent A_1B_1/A_1B_1. Gametes of this parent are all A_1B_1, while gametes of the hybrid are of four types, A_1B_1, A_1B_2, A_2B_1 and A_2B_2, the frequencies of which give the frequencies of the genotypes of the progeny (see the table below). What about the genotype at QTL Q? All gametes from fixed parent A_1B_1/A_1B_1 have allele Q_1, while gametes of hybrid will have alleles Q_1 or Q_2, with probabilities that may be expressed given the genotype of the marker, as a function of the ratio r_1/r. The means of the genotypic classes, θ_i, may be expressed as a function of the genotypic values μ_{11} and μ_{12} of genotypes Q_1/Q_1 and Q_1/Q_2, respectively.

Genotype of marker locus	Genotype of QTL	Probability	Genotypic mean θ_i
A_1B_1/A_1B_1	Q_1/Q_1	1	$\theta_1 = \mu_{11}$
A_1B_1/A_1B_2	Q_1/Q_1	$1 - r_1/r$	$\theta_2 = (1 - r_1/r)\mu_{11} + (r_1/r)\mu_{12}$
	Q_1/Q_2	r_1/r	
A_1B_1/A_2B_1	Q_1/Q_1	r_1/r	$\theta_3 = (r_1/r)\mu_{11} + (1 - r_1/r)\mu_{12}$
	Q_1/Q_2	$1 - r_1/r$	
A_1B_1/A_2B_2	Q_1/Q_2	1	$\theta_4 = \mu_{12}$

When r_1 tends to 0, θ_2 tends to θ_1 and θ_3 to θ_4. Intuitively, the difference $\theta_3 - \theta_2$, relatively to difference $\theta_4 - \theta_1$, will give information about QTL position between the markers. Indeed, a simple calculation shows that:

$$r_1 = \frac{r}{2}\left(1 - \frac{\theta_3 - \theta_2}{\theta_4 - \theta_1}\right)$$

In a backcross, it is not possible, unless the reciprocal cross is available, to separately estimate a and d, since the difference $\mu_{12} - \mu_{11}$ is equal to $a \pm d$. But this is possible in an F_2 population, even though the calculus is slightly more complex. This interval mapping method is an alternative to the lod scores method for mapping QTLs and estimating their effects (see text).

$(m_{22} - m_{11})/2$, where m_{22} and m_{11} are the mean values of homozygous genotypes M_2M_2 and M_1M_1, respectively (since $r = 0$, the genotypes at the QTL are therefore respectively Q_2Q_2 and Q_1Q_1) (Fig. 39). The additive effect is also the slope of the line passing through the mean values of genotypes M_1M_1 and M_2M_2, when the genotypes are ordered according to the number of doses of one of the alleles (Fig. 37). Naturally, a may be negative. The sign

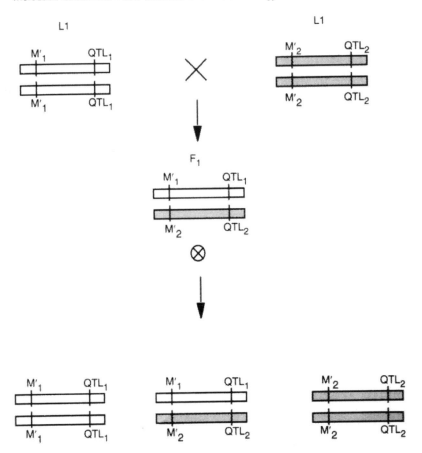

Figure 38. Interpretation of the difference between means with a model of a single QTL close to the marker. Here, only the cases in which there has been no recombination between the marker and the QTL are represented. QTL_1 and QTL_2 are the two alleles at the QTL.

of a indicates which parent carries the "favourable" allele, i.e., the allele that gives the highest value of the trait. With our notations, a positive value indicates that the favourable allele comes from parent 2; a negative value indicates that it comes from parent 1. Finally, it may be convenient to consider the effect of double substitution of one allele for another, $2a = m_{22} - m_{11}$.

The degree of dominance is the difference between the mean of heterozygotes M_1M_2 and half the sum of homozygotes: $d = m_{12} - (m_{11} + m_{22})/2$ (Fig. 39). It is more meaningful to use the ratio d/a, which allows the classification of the dominance. If $d/a = 0$, there is additivity (Fig. 37b). If $0 < |d/a| < 1$, there is partial dominance. If $|d/a| = 1$, there is complete dominance (Fig. 37c). Finally, if $|d/a| > 1$, there is over-dominance.

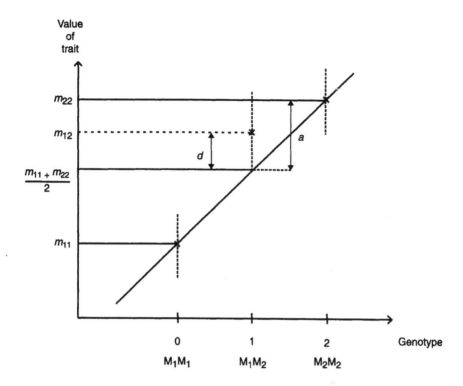

Figure 39. Parameters defining the action of a QTL. The genotypes at marker locus are ordered according to the number of doses of the allele M_2. The additive effect a is the slope of the line passing through the averages of homozygotes M_1M_1 and M_2M_2. It is thus equal to half the difference between these averages: $a = (m_{22} - m_{11})/2$. The degree of dominance is the deviation between the average of heterozygotes and the average of the two homozygous classes: $d = m_{12} - (m_{11} + m_{22})/2$.

In the general case, QTLs and markers are not totally linked. The consequence of recombination is that the individuals homozygous at the marker are not all homozygous at the linked QTL, because the other allele is present in a proportion that depends on the recombination rate between the marker and the QTL. In other words, the additive effect estimated is the additive effect *at marker locus a'*. The relation between a and a' depends only on the rate of recombination between the marker and the QTL: $a' = a(1 - 2r)$ (see box 4.3). For example, a QTL having an additive effect $a = 7.5$ g for a fruit weight leads to an *apparent* additive effect $a' = 5$g if it is located at 20.3 cM of the marker (with the Haldane distance).

MEASURES OF QTL EFFECTS

There are two common ways to define the "effect" of a QTL. First of all, as defined above, it pertains to *a difference of effects between alleles*, which can be

Box 4.3: Fine mapping of QTL

The population sizes conventionally used do not allow a precise location of QTLs with moderate effect. The confidence interval of the position may exceed 15 or 20 cM. Such segments may comprise several hundreds of genes, so any attempt at characterization or positional cloning of QTLs is impracticable. Moreover, the co-location of QTLs controlling different traits may be due to fortuitous linkage or to the pleiotropic effects of a single QTL. Fine mapping methods have been proposed that rely on recombination to split the genome into smaller fragments (Paterson et al., 1990; Eshed and Zamir, 1995; Liu et al., 1996). The two main genetic materials used are the near-isogenic lines and the highly recombinant inbred lines.

Near-isogenic lines

The simplest way to proceed is to use backcross to develop a first pair of near-isogenic lines for a region containing the QTL of interest. From their F_1 hybrid, an F_2 or backcross population is derived. Recombination produces individuals that are heterozygotes for *fragments* of the region containing the QTL. Markers will allow those individuals to be screened, and then their self-fertilization will allow individuals homozygous for the fragment to be obtained. By comparison of these new lines with the initial recurrent line, the QTL can be located more precisely. The figure below shows four lines that have fixed different fragments of a region of 12 cM containing a QTL (only one chromosome is represented, since the lines are homozygous). The fragments that come from the donor parent are in black, and the adjacent fragments, where a recombination has occurred, are in gray. Lines II and III have a significant difference (+) with the recurrent line for the trait studied (Case 1). The QTL is therefore in the interval of 2 cM between markers 4 and 5, since it is the only interval from the donor parent that is common to the two lines. This method also allows us to find out whether there are one or several QTLs in the region of interest. If, for another trait, the lines I and IV are found to be significantly different from the recurrent line (Case 2), it is concluded that there is a QTL between markers 1 and 3 (4 cM), and another

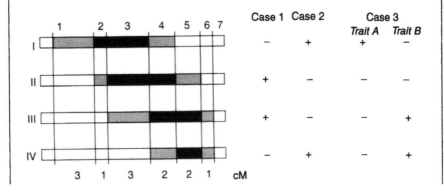

between markers 6 and 7 (1 cM). In some favourable cases more than two QTL may be detected (Eshed and Zamir, 1995). Finally, if two traits are being investigated and the goal of the experiment is to decide between genetic linkage and pleiotropy, linkage is concluded when the lines having a significant difference with the recurrent line are not the same for the two traits. In case 3, there is a QTL for trait A between markers 1 and 3, and a QTL for trait B between markers 5 and 7.

Highly recombinant inbred lines

Compared to the recombinant inbred lines, the *highly* recombinant inbred lines (HRIL) are developed by intercalating some generations of random mating between the F_2 progeny and the generations of self-fertilization, in order to increase the recombination. For two tightly linked loci A and B, and with n generations of random mating, we have the relationship $R_n/r \approx n+2$, where r is the recombination rate between A and B, and R_n is the *apparent* recombination rate in the HRILs. Therefore, the genetic map is linearly extended in proportion to the number of generations of random mating, as shown in the figure below:

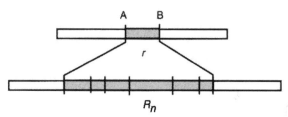

For example, with 8 generations of random mating, a distance of 1 cM (about 1% of recombination) will be 10-fold higher. In other words, and everything being equal, a QTL within an interval of 10 cM in a doubled haploid population will be mapped within 1 cM only with HRILs if $n = 8$.

The development of such genetic materials is of course time-consuming, and the number of markers in the regions of interest must be increased. But they may be reproduced identically by self-fertilization, thus making possible numerous studies in which QTLs will be directly mapped with great precision.

quantified as an additive effect (a or a') or as an effect of double substitution ($2a$ or $2a'$). For example, in a cross between the varieties of pea "Erygel" and "661", a marker located on the linkage group b had an apparent effect of 14 cm on the height, which is the difference in size between the means of individuals M_1M_1 and M_2M_2 at this marker. Similarly, on chromosome 4 a marker is found with an effect of double substitution of 3.4 for resistance to Ascochyta blight, a trait noted on a scale of 0 to 5 (Dirlewanger et al., 1994). The additive effect or the effect of double substitution is sometimes expressed in units of standard deviation: a/σ_T or $2a/\sigma_T$, where σ_T is the phenotypic standard deviation of the population.

On the other hand, each segregating QTL contributes to a part of the total phenotypic variation. This is quantified by the R^2, which is the ratio of the treatment (marker) sum of squares to the total sum of squares (see box 4.1). If the marker is not strictly linked to the QTL, the reduction of the difference between means reduces the treatment sum of squares and thus reduces the R^2 value. The sum of R^2 values of various independent QTLs for a trait cannot exceed 1 (100% of the variation explained), but in practice it is far from reaching this value, for several reasons. QTL with weak effects, potentially numerous, cannot be detected due to the limited power of the tests. Environmental effects account for a part of the total variation (the heritability of the trait is never equal to 1). Finally, a fraction of the variation may be due to epistatis between QTLs (see subsection *Epistasis*).

These two ways of accounting for the effect of a QTL are obviously not specific to their detection by analysis of variance in an F_2, and they remain valid no matter which detection method is used.

LIMITS OF QTL DETECTION ON INDIVIDUAL MARKERS

The advantage of QTL detection on single markers is simplicity. Any statistical software can be used for the analyses in large series. These analyses are robust to deviation from normality, and a saturated genetic map is not needed. Moreover, the models can easily be modified, for example to integrate a variable as cofactor, or to take into account a complex design. If the density of markers is low, however, this method is ineffective in estimating the actual position and effects of QTL. Other more powerful methods have thus been developed that allow us to map the QTL in the interval between two markers and to estimate precisely their effects at this position.

Detection of QTLs from two or more markers

METHODS OF INTERVAL MAPPING

Methods of interval mapping are based on the hypothesis that there is one QTL at most in the interval between two markers G and D linked with a recombination rate r. The genotypic value of each of the possible genotypes at two loci G and D (there are for example nine in an F_2) may be expressed as a function of a and d values of this QTL and of the rates of recombination between the QTL and its flanking markers. This system of equations can be resolved and provides the rates of recombination between the QTL and its markers (see box 4.3 for the backcross case), as well as the values a and d (Knapp et al., 1990). Another popular method for testing the presence of a QTL in an interval between two markers is based on the calculation of a LOD score. At each position of a chromosome (with a step of 2 cM for example), the decimal logarithm of the likelihood ratio is calculated:

$$LOD = \log_{10} \frac{V(a_1, d_1)}{V(a_0, d_0)}$$

where $V(a_1, d_1)$ is the value of the likelihood function under the hypothesis of presence of a QTL, where the estimates of the parameters are a_1 and d_1, and where $V(a_0, d_0)$ is the value of the likelihood function under the hypothesis of absence of QTL, that is, when $a_0 = 0$ and $d_0 = 0$ (Lander and Botstein, 1989). A LOD of 2 thus signifies that the presence of a QTL at a given point is 100 times more likely than its absence, a LOD of 3 means 1000 times more likely, etc. A curve of LOD can thus be traced as a function of the position on a linkage group. The maximum of the curve, if it goes beyond a certain threshold (see later), indicates the most probable position of the QTL (Fig. 40). The confidence interval of the position of the QTL is conventionally defined as the chromosomal fragment corresponding to a reduction of LOD of 1 unit below the maximum LOD, which indicates that the likelihood ratio has fallen by a factor of 10. A discussion on the confidence interval can be found in Mangin et al. (1994). With the most common sample sizes, around 100 to 200 individuals, and with QTLs having moderate effects, the confidence intervals are rarely smaller than about 20 cM.

There is a similar approach, equivalent in power, that consists of using tests and estimations of the linear regression model (Haley and Knott, 1992) in place of complex calculations of maximum likelihood. It has moreover the advantage of being easier to generalize for various types of experimental populations, and more flexible to use.

Comparison of the detection power of QTL between methods on individual markers and on flanking markers indicates that the latter is only

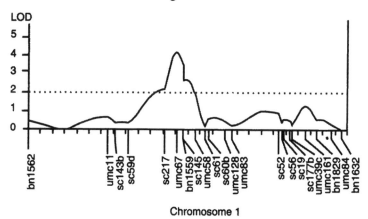

Figure 40. Example of LOD score curve obtained for female flowering time on chromosome 1 of maize. A population of recombinant lines was used. On the X-axis are the markers along the chromosome, on a scale of centimorgans, and on the Y-axis is the LOD value. The horizontal dotted line represents the threshold of detection (after Charcosset et al., 1994).

5% better than the former for intervals smaller than 20 cM. Beyond that, interval mapping has a rapidly increasing efficiency (Rebai et al., 1995).

The major limitation of interval mapping is that it does not allow precise detection of nearby QTLs on the same chromosome. Thus, when two QTLs located in small intervals act in the same direction (coupling), there is a risk of detecting a single "phantom" QTL between the two real ones (Martinez and Curnow, 1992). If the two QTLs are far enough away (more than 50 cM), it is easier to estimate the effect and the position of the second by fixing the position of the first, as suggested by Lander and Botstein (1989). However, this procedure is only effective if the QTLs are not located in adjacent intervals and if at least one has a relatively strong effect. An alternative method consists of searching for QTL in two dimensions by modelling the QTL pairs in various combinations of intervals of markers and by using multiple linear regression (Haley and Knott, 1992; Martinez and Curnow, 1992). This approach is nevertheless more cumbersome to use and less effective than classic interval mapping when a single QTL is present on the chromosome.

MULTIPLE MARKER METHODS

Kearsey and Hyne (1994) proposed an approach based on regression of additive effects at markers, a'_i, on the quantity $(1 - 2r_i)$, where r_i is the recombination rate between the marker i and the hypothetical QTL. At the position of the QTL there is the linear relation $a'_i = a(1 - 2r_i)$, the slope of which, a, is an estimation of the additive effect of the QTL (see box 4.3). For a given linkage group, this regression is performed every 2 cM, for example, the r_i corresponding to the rates of recombination between the hypothetical position of QTL and each of the markers i. The most probable position of QTL is that in which the residual of the regression model is the lowest. This method has the advantage of simplicity, but it has certain limitations. In particular, it cannot be used with dominant markers in F_2 populations.

Jansen (1993) proposed a combination of interval mapping and regression on markers, which would allow a more precise detection of multiple QTLs. In practice, this involves taking the other QTLs present in the genome, represented by their markers, as cofactors in the model, which reduces the part of residual variation induced by their segregation. The choice of the markers to be introduced in the model is an important point. Jansen and Stam (1994) proposed to take for each chromosome the most significant markers in a regression model step by step. One can also make a prior search for QTL by simple interval mapping and then take the markers closest to the QTLs thus detected. Zeng (1994) proposed a similar approach. Methods using cofactors are rapidly proliferating and sometimes seem to substantially improve the precision of estimation of QTL effects and positions.

Rodolphe and Lefort (1993) proposed a linear model approach by which chromosomal effects due to several linked QTLs can be detected globally with codominant markers. This method is interesting when the associated QTLs have very mild individual effects or opposite signs but is not optimal for detection of isolated QTLs.

Factors influencing QTL detection

Although the principle of QTL detection is relatively simple, several factors that influence the results must be taken into account to optimize the experimental designs.

DETECTION POWER AND WITHIN-CLASS VARIATION

In statistics, the type I error (α) is the probability of observing the effect of a factor when there is none, i.e., of observing an effect simply by chance. Applied to QTL detection, this corresponds to the probability of detecting a "false" QTL. The Type II error (β) corresponds to the probability of not observing the effect of a factor, when there is an effect. For QTL detection, this corresponds to the probability that the test does not reveal a QTL when there is one. The power of the test, $1 - \beta$, is thus the probability of making a good decision, in our case, of detecting a real QTL. Thus it is necessary to find a compromise to limit the number of false positives without reducing too much the power of the test. Besides, the choice of α and β will depend on the objective of the experiment. For a preliminary evaluation of "probable" positions of QTL, a high α (about 0.05 or 0.01) may be tolerated, while for initiating the cloning of a QTL a much lower value must be taken. At a given sample size, the power of QTL detection depends on the one hand on the additive effect of the QTL (a very small difference of effects between alleles will not be found significant) and on the other hand on the variance within the genotypic classes, or within-class variance. This variance depends on several factors:

- Environmental effects (the control of variations due to the environment increases the power of the test).
- Other segregating QTLs in the genome. Indeed the individuals of a genotypic class for a given QTL are not fixed for the other QTLs distributed on the rest of the genome and may thus present very different phenotypic values according to their genotype at other QTLs. Effects of epistasis may also aggravate the phenomenon. To improve the power of detection of mild-effect QTLs in a genetic background where a large-effect QTL has been found, a two-way analysis of variance model can be used, one of the factors being a marker of the large-effect QTL. By generalizing this principle, we find the multiple marker methods mentioned above (Jansen and Stam, 1994; Zeng, 1994). Moreover, an experimental solution consists of working on a new population in which the QTLs with a strong effect

have been fixed by self-fertilization (sub-section *Populations in which the genetic background is partly fixed*).

– The distance between marker and QTL.   As we have seen, recombination leads to the presence of both alleles of the QTL within each class of genotypes at the marker. This increases the within-class variance, which, associated with a reduction in the difference between averages, will reduce the power of the test as r increases. If a weak effect is found, one cannot distinguish between the existence of a QTL with strong effect but far from the marker and that of a QTL with weak effect closely linked to the marker (at least with methods on individual markers). Nevertheless, the fact that recombination affects the form of within-class distributions may help in distinguishing between these two, as has been shown by Asins and Carbonell (1988), Weller (1987), or Kearsey and Hyne (1994).

POPULATION SIZE AND MARKER DENSITY

Darvasi et al. (1993) showed that power of QTL detection and precision of QTL location depend more on population size than on marker density. Once a mean density of markers of 20 or 25 cM is attained, any supplementary means must be invested in analysing additional individuals rather than in increasing the number of markers. Figure 41 shows the influence of the population size studied on the power of QTL detection, as a function of their

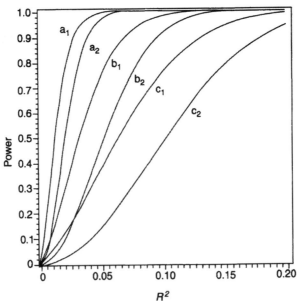

Figure 41. Power of detection of a QTL (Y-axis) as a function of R^2 (X-axis) in backcross, doubled haploid, or recombinant inbred line populations. a, b, and c correspond to populations of 500, 200, and 100 individuals, respectively. Subscripts 1 and 2 correspond to type I errors of 0.01 and 0.001, respectively (after Charcosset and Gallais, 1996).

effect. It is shown that a QTL with a large effect will be detected with a high probability no matter what the population size, but for detection of a QTL with moderate effect ($R^2 \approx 5\%$), it is necessary to use a larger number of individuals.

It must also be noted that for a given total number of individuals, it is better to increase the number of genotypes of the population than to increase the number of replications per genotype. A single replication on numerous genotypes allows the maximum power to be reached. Other data on test power as a function of size and type of populations studied are found in Knapp and Bridges (1990).

CHOICE OF STATISTICAL RISKS

In many experimental studies, type I errors chosen for analysis of variance on individual markers are $\alpha = 0.05$ or $\alpha = 0.01$. Because of the large number of analyses done, these high values inevitably lead to the detection of false QTLs (100 independent tests with $\alpha = 0.05$ produce on average five false positives). Even if the markers are not independent, lower values of α must thus be chosen according to the number of markers, while keeping in mind that this increases the type II error (i.e., decreases the power of the test). Similarly, for interval mapping, a *global* type I error of $\alpha = 0.05$ for the entire genome imposes a fairly high LOD per interval, which depends on the density of markers and the genetic length of the genome. Lander and Botstein (1989) proposed for example a threshold LOD of 2.4 for a genome of 1000 cM with a marker every 15 cM. Rebai et al. (1994) subsequently developed a method that allowed analytical calculation of an approximate threshold for each linkage group as a function of its length and marker density. Churchill and Doerge (1994) suggested permutation tests, based on a random resampling of data, to evaluate an empirical and robust threshold. Finally, it is noted that reducing the number of individuals analysed reduces the power of the tests, i.e., actual QTLs will be "lost", especially those with weak effect, but the QTLs detected are not less reliable. The risk that a detected QTL is a false positive is α, chosen by the scientist, and does not depend on the size of the population.

DEVIATION FROM NORMALITY

The method of interval mapping based on LOD score is applied to traits with continuous variation, presenting a normal distribution of residuals. For traits measured on a scale, such as disease resistance (often noted from 0 to 5), and/or those with a highly asymmetric distribution, the use of this method risks giving aberrant results. One can thus use non-parametric tests that require no hypothesis about the distribution of data (Kruglyak and Lander, 1995) or specific methods for traits in classes (Hackett and Weller, 1995). Rebai (1997) showed nevertheless that if the population studied is large enough, the test of interval mapping based on regression is as effective as non-parametric tests, and this is true for a large range of distributions.

A brief description of QTL detection software's is given in box 4.4.

Box 4.4: Major software's for QTL detection

- Mapmaker/QTL (Lincoln, Daly and Lander) is the program provided with the mapping program Mapmaker/Exp and is widely used. It uses the method of maximum likelihood by interval mapping, and treats F_2, doubled haploid, and backcross. It is free and functions on most systems.

- QTL Cartographer (Basten, Weir and Zeng) searches for QTLs by regression marker by marker, interval mapping (method of Lander and Botstein, 1989), or composite interval mapping (Zeng, 1993), for various population types. It functions on MS-Windows, MacIntosh, and Unix.

- Qgene (Nelson) functions on MacIntosh. It searches for QTLs by regression marker by marker or by regression on interval (method of Martinez and Curnow, 1992). It has a very user-friendly graphic interface that allows exploration of molecular and phenotypic data. All types of populations resulting from crosses between homozygous lines are analysed.

- MapQTL (van Ooijen) works along with Joinmap and uses interval mapping (method of Lander and Botstein, 1989), multiple QTL mapping (MQM, Jansen, 1993), and non-parametric tests for traits with discrete variation. It treats different types of populations, including F_1. It functions on MacIntosh, MS-DOS, and Unix and costs around US$ 1500.

- PLABQTL (Utz and Melchinger) searches for QTLs in F_2, recombinant inbred lines, or doubled haploid populations by simple or composite interval mapping and by multiple regression. It functions on MS-DOS.

- MQTL (Tinker and Mather) works on Unix and finds QTLs in experimental designs with several environments by composite interval mapping.

- MultiCrossQTL (Rebai and Jourjon) searches QTL on a large range of populations and allows simultaneous treatment of several connected populations derived from diallelic crosses. It relies on a generalization of interval mapping methods by regression and on that of multiple QTLs of Jansen (1993). It works on Unix and MS-DOS.

- MultiQTL, in addition to the classical interval mapping analysis, allows the simultaneous search for QTLs using multiple trait information and several environments (Korol et al., 2001).

Access to most of these software's is free and the addresses of sources can be found in databases including http://www.stat.wisc.edu/biosci/linkage.html. This server also provides information on other programs used in human genetics and enables simulations of markers and phenotypic traits.

Advantages and disadvantages of populations currently used for QTL detection

COMPARISON OF F_2, BACKCROSS, DOUBLED HAPLOIDS, AND RECOMBINANT INBRED LINES

With backcross, doubled haploids, or recombinant inbred lines, unlike F_2, there are only two averages to be compared for a given marker, which can be done by means of Student's t-test or a more sophisticated method. But all the effects cannot be estimated from these populations. In doubled haploids and recombinant inbred lines, a can be estimated, but not dominance. In backcrosses, where a homozygous genotype is compared to a heterozygous one, the effect estimated $(m_{12} - m_{11})$ or $(m_{12} - m_{22})$ will equal $\pm a$ or $(\pm a')$ only under the hypothesis of absence of dominance. Otherwise, the value found depends on two parameters, since it is equal to $\pm (a + d)$ or $\pm (a - d)$, without any possibility of estimating these separately (except using reciprocal backcrosses). In case of complete dominance, where $|d| = |a|$, a QTL may be detected only if the dominant allele is not the allele fixed in the recurrent parent. In recombinant inbred lines, recombination between the marker and the QTL will reduce the estimated additive effect more rapidly than with other types of population. By a reasoning similar to that described in box 4.3, it can indeed be shown that $a' = a(1 - 2R)$, where $R = 2r/(1 + 2r)$ (Chapter 2). Thus, at fixed r and a, the ratio of additive effects estimated at the marker between recombinant inbred lines and F_2 will be $1/(1 + 2r)$. For example, if $r = 0.1$, the additive effect at the marker in recombinant inbred lines will be 0.8 times that of other populations.

Around half the individuals of an F_2 population are heterozygous at a given locus and therefore are not informative for estimation of a. At equal population size, the power of tests from doubled haploid populations or recombinant inbred lines should thus be clearly superior. Still, this gain is partly (but only partly) compensated by the fact that the within-class variation is higher for the two latter types of populations because of the absence of heterozygotes at other QTLs. For the recombinant inbred lines, the more rapid decrease of the additive effect with r indicated above reduces the power of the tests; however, several individuals can be analysed for each genotype, as is the case in doubled haploids.

If there is dominance at a marker locus, the mix of individuals homozygous for one of the alleles with heterozygotes will reduce the power of the test for QTL detection. Besides, it will prevent the estimation of genetic parameters, including dominance at QTL. Finally, when the phenotypic value of the F_2 individuals is estimated by the average of their F_3 offspring, as is often done, the degree of dominance estimated will be half the real degree of dominance, because the progeny of a heterozygous F_2 genotype comprises only 50% heterozygotes.

Progeny derived from heterozygous parents

When no homozygous parental lines are available (in strictly allogamous species and species with a long generation time, such as trees), QTL detection is complicated because the parents may differ by more than two alleles, and because the phase (coupling or repulsion) of the marker-QTL linkage may be different from one family to another. Various progeny may nevertheless be used. Knowledge of the grandparent genotypes at marker loci can improve detection by allowing phases of associations between adjacent markers to be identified. It is this type of population with three generations that was used by Bradshaw and Stettler (1995) to locate QTLs for wood density in poplar. In eucalyptus, Grattapaglia et al. (1995) used information from two generations in families of full siblings to study ability for vegetative multiplication. The problem of QTL detection in F_1 populations derived from non-fixed parents was treated by Leonards-Schippers et al. (1994).

Populations in which the genetic background is partly fixed

The power of detection of a particular QTL in a given genetic background is low to the extent that other QTLs are numerous and/or have a strong effect. The populations conventionally used are made up of individuals in which the genomes comprise fragments of the two parental genomes in equal parts (statistically). This situation is the least favourable for detection of QTL with weak effect, since it maximizes the number of QTLs in segregation and possible epistatic interactions, and therefore results in large within-class variance. For this reason, Eshed et al. (1992) created introgression lines in tomato. By successive backcrosses and selection on the basis of RFLP markers, they obtained about 50 lines of the cultivated species *Lycopersicon esculentum*, the genome of each containing a chromosomal segment (of 33 cM on average) of the wild species *L. pennellii*. This population thus constitutes a kind of "genome library" of *L. pennellii* in the genome of *L. esculentum*. These lines were compared with the parental cultivated line to search for QTL on introduced fragments more efficiently than in a classical cross, because of the fixation of the rest of the genome. This was effectively observed. Twenty-three QTLs for dry matter content in fruit and 19 for fruit weight were detected (Eshed and Zamir, 1995), while in a population of recombinant inbred lines derived from the same cross, Goldman et al. (1995) detected only 7 and 13 QTLs, respectively, for the same traits.

This material also served as starting point for experiments of fine mapping (see box 4.2). By deriving an F_2 population from crosses between an introgression line and a cultivated line, then self-fertilizing the individuals carrying a recombination in the fragment, fixed lines for different subgroups of the initial fragment were constituted. For example, on the long arm of chromosome 2, where a QTL for fruit weight was found on a fragment of 55 cM, three QTLs were detected, in regions of 3.2 cM, 3.7 cM, and 14.1 cM, respectively (Eshed and Zamir, 1995).

Another way to have a genetic background partly fixed is to search for QTLs in so-called "advanced backcrosses" (BC2, BC3, BC4) (Tanksley and Nelson, 1996). This strategy has been successfully used for screening positive alleles from a wild related species, and it allows some generations to be saved for the production of improved lines since QTL detection and selection of genotypes are done in the same generation (see also the section *Transgression*).

Selective genotyping

Greater test power and significant economy of work can be achieved by molecular genotyping exclusively those individuals showing the extreme values of the trait studied (selective genotyping; Lander and Botstein, 1989). Such individuals have indeed high chances of accumulating either many favourable alleles or many unfavourable alleles for the various QTLs of the trait studied, so that, for each marker linked to a QTL, there will be simultaneously a greater difference between averages and a smaller within-class variation. As phenotypic data are generally less difficult to obtain than mapping data, one can work on large populations and thus have highly contrasted extreme classes. But this approach is useless when several traits must be measured, since there is no reason for the same individuals to be retained in the extreme classes (except when there are strong correlations between traits). On the other hand, the additive effects will be overestimated because the genotypes at various QTLs are no longer independent: individuals of high phenotypic value have fixed the favourable alleles at most of their QTLs, and those of small value have fixed the unfavourable alleles (Darvasi and Soller, 1992).

Bulked segregant analysis (BSA), which was proposed for mapping major genes (see Chapter 3), can also be used here for traits with high heritability and QTLs with large effects (Chalmers et al., 1993; Wang and Paterson, 1994). The markers found to be correlated to the trait may correspond to several non-linked QTLs, which must be verified by mapping (see, for example, Timmerman-Vaughan et al., 1996). Using BSA, Tanhuanpaa et al. (1996) detected several RAPD markers very closely linked to a QTL for oleic acid content in rapeseed. Mansur et al. (1993a, b) used this technique to verify the validity of QTL for yield components detected in an F_2 population of 69 individuals, analysing the extremes of a population of 280 recombinant inbred lines. For this purpose they hybridized the RFLP probes that revealed QTLs in F_2 with the mix of DNAs of individuals of extreme phenotypes, and thus they were able to confirm the primary linkages. Darvasi and Soller (1994) showed that this approach can be made much more efficient if the allelic frequencies in each group can be quantified.

GENETIC AND MOLECULAR BASES OF QUANTITATIVE TRAIT VARIATION

Genetic data on QTLs

QTLs ARE (ALMOST) ALWAYS DETECTED

Ever since the mapping of QTL became possible, several studies showed that even with populations of moderate size (sometimes hardly more than 50 individuals), some QTLs are almost always found, for all types of traits (Tanksley, 1993; Kearsey and Farquhar, 1998). This means that, in many cases, a large part of the variation can be explained by the segregation of a few QTLs with major effects. It is not rare to find single QTLs that explain 20% of the variation, or even more. And as the R^2 takes generally into account the environmental variation, the genetic variation attributable to QTLs is actually greater. Data compiled on maize, cereals, brassicas, tomato, and other plants in which many QTLs have been mapped indicate that the QTL effects measured by their R^2 are distributed according to a L curve, with a few QTLs having strong or very strong effect, and more QTLs having moderate or weak effect (Fig. 42) (Kearsey and Farquhar, 1998). Bost et al. (1999, 2001) showed that biochemical, genetic, and statistical factors may explain such a distribution.

With populations of the common size (60 to 400 individuals), and depending on the traits, one to ten QTLs are commonly detected (Tanksley, 1993; Kearsey and Farquhar, 1998; see also http://ars-genome.cornell.edu). These numbers constitute a minimum estimate of the QTLs in segregation in the populations studied, for several reasons:

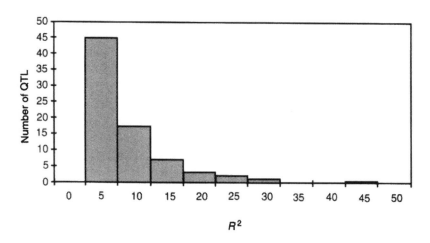

Figure 42. Schematic histogram illustrating the L-curve of the distribution of QTL effects as revealed from numerous studies, compiling data over various traits.

- Some QTLs have an effect below the detection threshold. When larger populations are studied, QTLs with a weak effect are effectively detected. In an F_2 of 1700 individuals, Edwards et al. (1987) detected some QTLs explaining less than 1% of the total phenotypic variation.
- The effect of a chromosomal segment may be due to several linked QTLs.
- If two QTLs of comparable effect are closely linked, a QTL will be detected only in coupling phase, i.e., if the positive alleles at the two loci come from the same parent.

Moreover, as always in genetics, the monomorphic loci in a given population cannot be detected. Thus, the series of QTLs that contributes to the variation of a trait does not correspond to the series of genes that must be functional for this trait to be expressed. If the trait considered is, for example, the concentration of a final product of a linear metabolic pathway, *all* the genes of the pathway must be functional for the product to be present, but only the *subgroup of those that are polymorphic* in the population studied can possibly influence the variation of this trait. It is moreover commonly observed that the series of QTLs for a given trait in two different populations only partly overlap (Tanksley and Hewitt, 1988; Beavis et al., 1991). In addition to the possible effects of epistasis (see section *Epistasis* below), the most immediate explanation is that it is not exactly the same loci that are polymorphic in the two populations.

DOMINANCE EFFECTS

Even though tests of significance are rarely performed, dominance effects seem to be found quite often. In some cases, the value $|d/a|$ clearly goes beyond 1, as there was overdominance for the QTL detected (Stuber et al., 1992). In another interpretation, however, it is supposed that one is dealing with not one, but two linked QTLs in repulsion, with dominance for each one. The double heterozygotes $+ -/- +$ will thus have a higher phenotypic value than each of the double homozygotes $+ -/+ -$ and $- +/- +$. For example, thanks to a fine mapping experiment using near-isogenic lines, the origin of the apparent overdominance detected in a region of 37 cM for fruit content in soluble solids in tomato could be interpreted as the consequence of a linkage between two dominant QTLs in repulsion (Eshed and Zamir, 1995). If fine mapping does not allow us to decide between the two hypotheses, positional cloning of the locus or loci may prove to be necessary.

TRANSGRESSION

There is *transgression* when individuals of progeny go beyond the range of parental variation for a given trait. This phenomenon can be genetically analysed through QTL mapping. The transgression may be explained on the basis of a simple model with only two independent loci. If the parents have respectively the genotypes $+-/+-$ and $-+/-+$, their progeny will contain individuals $++/++$ and $--/--$, the genotypic values of which will be above and below the parental values. By generalizing this model to several loci, it is expected that the more the parents differ for the trait considered, the less we find transgressive individuals among the offspring, and the more the parents will be genetically homogeneous at the various QTLs detected (i.e., it is expected that one has fixed "favourable" alleles for most of the QTLs, the other the "unfavourable" alleles). In a study on many traits in tomato, De Vicente and Tanksley (1993) were able to confirm this prediction. On the contrary, transgressions and QTLs could also be found in a population whose parents do not present any difference for the trait studied, simply because a given value of the trait can be obtained with different genotypes at the QTLs of the trait (this is notably what can be expected with unrelated individuals).

It is interesting to note that even when highly contrasted individuals have been chosen as parents of a population, it is not rare to find a fraction of QTLs showing an effect inverse of that expected from the value of the parents. In a search for a QTL controlling resistance to Ascochyta blight in pea, Dirlewanger et al. (1994) observed that the resistant parental variety had fixed the allele for susceptibility for one out of three QTLs detected (the one showing the weakest effect). The resistance of this line could thus be improved further by introgressing the resistance allele found in the susceptible parental line. This type of construction has effectively been done in tomato, from interspecific crosses between the cultivated species *Lycopersicon esculentum* and wild related species *L. hirsutum* and *L. pimpinellifolium*. Advanced backcross (BC_3) analyses allowed QTLs for traits of agronomic importance to be mapped (Bernacchi et al., 1998a), then the development of NILs carrying introgressions for desirable wild QTL-alleles revealed that for most traits the predicted phenotypic improvement was observed (Bernacchi et al., 1998b).

Epistasis

A search for epistasis between QTLs has sometimes been done. By n-way analyses of variance with interactions, one can in principle look for interactions of any order between QTLs. But practically this approach is limited to interactions between pairs of QTLs, because the increase in the number of loci quickly reduces the number of individuals per genotypic class. With interactions of the second order (between three QTLs), there are 27 genotypic classes in an F_2, thus very few individuals per class and a very low power of the test. Moreover, even taking into account interactions of the first order, a large number of two-way analyses of variance must be done. With N loci, there will be $N(N-1)/2$ tests, which poses the problem of the choice of type I error. Despite these difficulties, epistatic interactions have been found in various studies. In some cases very clear interactions between two QTLs have been observed, for example for tolerance to low phosphorus stress in maize (Reiter et al., 1991), salt tolerance in tomato (Breto et al., 1994), hardening to cold in barley (Pan et al., 1994), architecture of maize (Doebley et al., 1995), components of resistance to *Phytophthora capsici* in pepper (Lefebvre and Palloix, 1996), or even molecular traits such as protein concentration (Damerval et al., 1994).

Nevertheless, various authors have underlined the infrequency of epistasis (Paterson et al., 1988; Stuber et al., 1992; Schon et al., 1994, etc.; see discussion in Tanksley, 1993). This result apparently contradicts both numerous results in classical quantitative genetics, which frequently find variance of epistasis, and theory of metabolic processes, in which epistasis appears as an intrinsic property (Kacser and Burns, 1981). As various authors have emphasized, the principal reason for this is undoubtedly statistical: the populations used are often of modest size, and the analysis of variance has a limited power for detecting interactions. A way of increasing the power is to eliminate the "background noise" due to other QTLs, using near-isogenic lines for the QTLs of interest as parents of the populations studied. For example, in a semi-diallele arrangement, Eshed and Zamir (1995) crossed 10 introgression lines of tomato, each having a specific fragment of genome of *L. pennellii* (see above) in the genetic background of the cultivated tomato *Lycopersicon esculentum*. They thus proved numerous significant epistatic interactions. Similarly, in order to test for epistasis among QTLs involved in the evolution of maize from its ancestor (teosinte), teosinte alleles of two QTLs previously shown to control much of the morphological difference between these plants were introgressed into a genetic maize background. Plants of each of the four two-locus homozygous classes of the two QTLs were compared, and strong interaction effects were observed on morphology and on the level of the mRNA of one of the genes involved in the maize evolution (*tb1*) (Lukens and Doebley, 1999).

On the other hand, it is not because a QTL does not show epistatic interactions with other QTLs taken individually that its effect is independent of the genetic background. As shown by Doebley et al. (1995), the effects of two QTLs of maize domestication are much weaker when they are segregating in a "teosinte" genetic background than in an F_2 maize × teosinte background. Similarly, in studying three populations of recombinant inbred lines of maize resulting from three possible crosses between three lines, Charcosset et al. (1994) were able to show the significant influence of genetic background on the expression of several QTLs for flowering time and yield.

GENETIC ANALYSIS OF CORRELATION BETWEEN TRAITS

QTL mapping provides access to the genetic bases of correlations between traits. When traits are correlated, it is expected that at least some of their QTLs are common (or at least genetically linked). This is effectively what is observed. For example, in the F_2 population of pea already mentioned, the strongest QTLs for height, flowering time and number of nodes, which are highly correlated traits, were found in a single region (Dirlewanger et al., 1994). The same has been observed in barley for height and flowering time (Bezant et al., 1996), with in addition two QTLs specific to height and seven to flowering time. In maize under low phosphorus stress, where growth rates of shoot and root of the young plantlet are correlated, the search for QTLs showed that four chromosomal regions out of six identified were involved in the control of both traits, each of the other two QTL controlling only one of the traits (Reiter et al., 1991). This possibility of distinguishing the common QTLs from the specific ones can prove useful in selection. If two favourable traits are negatively correlated, the discovery of specific QTLs allows us to estimate the extent to which each of the traits can be improved without negative consequence for the other trait. In this connection, Korol et al. (1995, 2001) proposed a statistical test to use the information of correlated traits to locate QTLs simultaneously controlling several traits. They showed that this approach increased the power of QTL detection when compared to the search trait by trait.

In case of apparent co-location of QTLs controlling different traits, there is no direct method to decide between the presence of a single QTL with pleiotropic effect or of two linked QTLs. In the latter case, fine mapping may give the clue (see box 4.2). Nevertheless, there is no symmetry of the evidence: the impossibility of finding two QTLs does not prove there is only one, but may simply indicate that the linkage is close. Apart from the implementation of cumbersome methods of molecular biology, one could rely on indirect arguments:

- If a mutation is known to jointly affect both traits, the pleiotropic relation is probable. If in addition the mutation and the QTL are co-located, the mutant locus represents a good QTL candidate.
- If both traits share not only one but two or more apparently common QTLs, the probability of pleiotropy is reinforced. Indeed, if a coincidence of position is not unlikely because of the size of confidence intervals of QTL positions, multiple coincidences are hardly likely to be due to chance.
- Physiological and/or morphological considerations may be put forward. If, for example, a common QTL is found for abscisic acid content and stomatal conductance in a maize progeny subjected to water stress, the causal relation is likely (see discussion in Prioul et al., 1997).
- The evolution of the marker-trait association over several generations will be stable if there is pleiotropy, while it will decline in case of linkage, because of recombination. Nevertheless, in case of close linkage the decrease may be very slow.
- Finally, in case of pleiotropy the correlation must be observed even in the absence of linkage disequilibrium, i.e., in unrelated individuals.

GENETIC DISSECTION OF TRAITS

QTL mapping is a particularly interesting approach in attempting to analyse the genetic bases of complex traits, focusing on *components* of these traits, which are assumed to have higher heritability. For example, the grain yield of maize may be analysed via a series of characteristics of the ear (Stuber et al., 1987), the plant height via the number of nodes, length of internodes, height at the ear (Edwards et al., 1992), etc. Veldboom and Lee (1994) studied the grain yield of maize and six of its components (number of ears per plant, mean length and diameter of the ear, mean number of rows, etc.). Out of 14 regions carrying QTLs, nine affected more than one component and two controled four and five components, respectively (with various correlation values between traits). One can also attempt to examine the role of physiological components in the expression of developmental or morphological traits. Causse et al. (1995) observed cases of apparent co-locations of QTLs for activities of carbon metabolism enzymes and QTLs for growth in maize. Similarly, taking components into account in the study of disease resistance allows us to refine our understanding of the mechanisms involved. In maize infected by an attack of the fungus *Exerohilum turcicum*, Freymark et al. (1993) separately studied the size and number of lesions, instead of giving an overall notation of infection. They showed that these two components are under independent genetic control, since their QTLs (two and three respectively) did not coincide. Lefebvre and Palloix (1996) also showed that resistance to *Phytophthora capsici* in pepper, studied using six components corresponding to different reactions of organs and stages of infection, was controlled by several QTLs, some specific to one component, others common to several.

EFFECT OF THE ENVIRONMENT

The environment is a factor that may have a significant impact on QTL effects. A corollary of the fact that the genotype-environment interactions may partly modify the rank of genotypes is that the amplitude of QTL effects may differ from one environment to another, and even that QTLs detected in one environment may no longer be found in another. This has been frequently observed, even though environmental influence differs according to the traits, range of environments chosen, and size of population (Paterson et al., 1991; Koester et al., 1993; Stuber et al., 1992; Schon et al., 1993; Bubeck et al., 1993; Hayes et al., 1993). An important concept on which these studies are based is that certain QTL are detected in all or almost all the environments tested and are "generalist", while others are found in a single environment and are "specific". This distinction must be taken into account in marker-assisted selection, according to the type and range of environments for which the varieties are developed.

In methodological terms, most of the studies simply look for QTLs in each environment and then compare the QTLs detected. Other studies use models of analysis of variance or multiple regression including, in addition to QTL effects and environmental effects, QTL-environment interactions (Zehr et al., 1992; Hayes et al., 1993; Jansen et al., 1995; Beavis and Keim, 1996). Software's developed by Zeng (1993) and Korol et al. (2001) can be used to test this interaction. In a more sophisticated method, Romagosa et al. (1996) took up again the data of Hayes et al. (1993) on grain yield in barley in 16 environments. They first did a principal component analysis on the matrix of genotype × environment interactions and then looked for QTL for principal components of this interaction. Four chromosomal segments were thus detected, which accounted for most of the differences in various environments.

COMPARATIVE QTL MAPPING

Using comparative mapping data (Chapter 2), some studies attempted to find out whether QTLs of a given trait are found at homologous positions on the genomes of related species. The most convincing example was found in the genus *Vigna* (Fabaceae), where QTLs for seed weight were mapped in the genome of the species *V. unguiculata* and *V. radiata* (Fatokun et al., 1992). On the basis of markers used, the regions carrying the QTL with the largest effect in each species (R^2 of 36.5% and 32.5%, respectively) appeared homologous. An additional argument in favour of the identity of these QTLs is provided by epistasis. The independent QTLs with which these major QTLs have an epistatic interaction are also located in homologous regions. Besides, the use of molecular probes general to Fabaceae revealed that QTLs for grain weight in peas (Timmerman-Vaughan et al., 1996) and soyabean (Maughan et al., 1996a) could correspond to one or another of the QTLs of *Vigna*. A compilation of QTLs for domestication traits in rice, maize, sorghum, and sugarcane also revealed coincidences of QTLs more frequently than might be expected purely by chance, especially for grain size, shattering, and flowering time in short days (Paterson et al., 1995). Finally, analogous results were published for three species of the genus *Lycopersicon*, including cultivated tomato (Paterson et al., 1991; Goldman et al., 1995).

These results are not only interesting as supplementary examples of conservation of synteny. The remarkable fact is that species that diverged very long ago (the Poaceae date back to around 60 million years) and live in often highly diverse ecological conditions display still homologous polymorphic loci explaining the variation of certain traits. At the intra-specific level, let us recall that comparison of two progeny may lead to partly different series of QTLs.

Characterizing QTLs

Characterizing QTLs is a difficult task, due to their weak phenotypic effects, and to the background noise caused by other QTLs and by environmental effects. In plants, very few QTLs have been definitively characterized at the molecular level. It is clear that knowledge of the nature of QTLs will be of great fundamental and practical interest. "Characterizing" a QTL has actually two different meanings. If nothing is known about the physiological and molecular bases of the trait, genetic and molecular techniques will have to be used to clone the QTL in question ("positional cloning"). On the other hand, if known-function genes are known to be involved in the expression of the trait, it will be more direct to search for those whose polymorphism accounts for a fraction of the variation of the trait (the "candidate genes" approach).

QTL CLONING

For isolation of QTL by positional cloning, the genetic techniques described in Chapter 3 can be considered. However, the continuous variation of the traits makes the task much more complicated. First of all, for statistical reasons, the size of the experimental designs must be larger. Then, for the constitution of near-isogenic lines or for BSA, markers flanking the QTL must be used to screen the genotypes, because the phenotypic screening is usually not possible. High-resolution mapping in the QTL region is therefore required. Positional cloning can really be considered when the QTL is located in an interval smaller than one centimorgan, and only in species with a small genome. The ideal is to attain a distance between markers and QTL of around the size of a YAC or BAC clone, or even less. Transformation may prove to be necessary to identify the QTL with no ambiguity, but other evidence has proved to be effective, such as that mentioned in Chapter 3 for cloning of major genes: e.g., high resolution mapping of single candidate genes to look for the clones that cosegregate with the trait, bioinformatic analysis of the sequence, or studies of expression (see below). In any case, QTL identification by cloning is expected to be possible, or at least easier, with large-effect QTLs, and/or for traits for which mutants are available (e.g., cloning dwarf genes to characterize QTL for height).

In spite of these drawbacks, some examples of successful positional cloning of QTLs have been recently published. In tomato, the QTL *fw2.2* is largely responsible for the dramatic increase of fruit size during domestication. High-resolution mapping allowed a YAC containing this QTL to be identified. When transformed into large-fruited cultivars, a cosmid derived from the *fw2.2* region of a small-fruited wild species, and containing one of the five genes found in the YAC (*ORFX*), reduced the fruit size by the predicted amount (Frary et al., 2000). Also in tomato, a QTL increasing sugar content in fruit was accurately mapped in a progeny with

7000 F$_2$ plants. Using near-isogenic lines for the QTL, high-resolution mapping delimited the QTL in a 484 bp region spanning an exon and intron of an invertase gene, an enzyme that modulates sink strength of tomato fruit (Fridman et al., 2000). In rice, a major quantitative trait controlling response to photoperiod, *Hd1*, was identified by means of high-resolution mapping using 1505 segregants, which enabled the authors to define a genomic region of only 12 kb as a candidate for *Hd1*. Further analysis revealed that *Hd1* corresponds to a gene that is homologous to *Constans* in *Arabidopsis* (Yano et al., 2000). Finally, in *Arabidopsis*, a QTL for flowering time was recently cloned using similar methods (El-Din El-Assal et al., 2001).

CANDIDATE GENES

The "candidate gene" approach is actually two different approaches (Table 13), which sometimes converge.

Table 13. The candidate gene approaches.

"Functional" candidate genes	"Positional" candidate genes
Known-function gene	Genetic mapping of a QTL
⇓	⇓
Polymorphism?	Genes/mutations in the region?
⇓	⇓
Does this polymorphism account for a part of the variation of the trait?	Does one of them correspond to the QTL?

"Functional" candidate genes

For some quantitative traits, the physiology indicates the functions involved. If the corresponding genes are available, whether they are QTLs of the trait studied or not depends on the effects of their polymorphism, if any, on the variation of the trait (Pflieger et al., 2001).

As far as we know, the first published example in plants of a known-function gene that could correspond to a QTL is *Shrunken-2*, which codes for one of the subunits of ADP-glucose pyrophosphorylase (ADPGppase), a key enzyme in the synthesis of starch in the grain. In a population of about 100 F$_3$ families of maize, Goldman et al. (1993) found that a QTL with a very strong effect for starch content was located precisely in the region of this gene. Another example is that of sucrose synthase, responsible for hydrolysis of sucrose into glucose and fructose. In a population of recombinant inbred lines of maize, QTLs for the activity of this enzyme were detected, one of which was located precisely at the level of its structural gene *Sh1*, on chromosome 9 (Causse et al., 1995). Moreover, a QTL for hexose content and one for early growth were also found in the same region. Polymorphism of *Sh1* could thus cause the variation in activity of the enzyme, which in turn would explain the presence of a QTL for products of enzyme reaction, and the influence on growth.

Another example of QTL characterization through the functional candidate gene approach is that of maysin content in maize silks. Maysin is a flavonoid that inhibits growth of a corn earworm (*Helicoverpa zea* Boddie). The pathway of its biosynthesis is well characterized, and the genes coding for the major enzymes and most of the regulatory factors were cloned. Using genomic and cDNA probes of these genes as markers for the search for QTL for maysin content in an F_2 population, the major role of *p1*, a transcription activator, was revealed (R^2 of 58%) (Byrne et al., 1996). This QTL acts in an additive fashion, with a great difference of effects between alleles: the averages of classes Q_1Q_1, Q_1Q_2, and Q_2Q_2 are respectively 0.023%, 0.280%, and 0.630% of fresh weight of silks. The large role of this regulator gene compared to enzyme coding genes could be fortuitous, but it is consistent with the metabolic control theory (Kacser and Burns, 1981), which shows that variations in efficiency of structural genes of enzymes in complex metabolic pathways generally have mild consequences as compared to variations in regulatory factors. This experiment also illustrates an important point, i.e., that the question of QTL characterization is not necessarily the same as the question of cloning of *new* genes. Even if we had all the genes and their function in a model species, the question of their impact on quantitative variation would remain.

Positional candidate genes

For many traits, there are no physiological data to guide the choice of known-function genes. The strategy is then to map the QTL as accurately as possible, and to search in the region of the QTL for genes or mutations that could turn out to be good candidates.

Of particular interest are mutations that confer an extreme value to the trait studied. For example, a QTL of chromosome 1 of maize, which affects the branch length in progeny between maize and teosinte, is coincident with *Teosinte branched 1* (*Tb1*), the mutation of which gives to maize a phenotype partly similar to that of teosinte, i.e., with long lateral branches. Subsequent cloning of *Tb1*, along with genetic and molecular studies (see below), confirmed that *Tb1* locus corresponded to the QTL, for which three types of alleles were found: a fully active allele in maize, an allele called *tb1 + teosinte* about half as active as *Tb1* in teosinte, and an inactive allele, *tb1*, in maize mutants.

Model species with small genomes entirely sequenced are particularly useful in the context of the candidate gene approach. The example of the *CO* (*Constans*) gene of *Arabidopsis* is illustrative. The mutant (*co*) delays flowering time in long photoperiod but not in short photoperiod and reduces the response to vernalization. The *CO* probe, as well as about 10 probes of markers linked to *CO* in a linkage group of 7.5 cM, have been mapped by RFLP in a backcross population of *Brassica nigra*. Because of duplications that occurred in the *B. nigra* genome, the linkage group was found in three distinct regions, two of which carried QTLs for flowering time coincident with homologues of *CO*. These results thus designate the homologue of *CO* of *Arabidopsis* as a good candidate for flowering time in *B. nigra* (Lagercrantz et al., 1996). It is interesting to note that *Hd1*, the major photoperiod sensitivity QTL in rice mentioned above, is closely related to the *Arabidopsis* gene *Constans* (Yano et al., 2000).

VALIDATION OF CANDIDATE GENES

Whether the positional cloning or candidate gene approach is used, the confirmation of the role of a gene in the variation of a quantitative trait may rely on different non-exclusive strategies.

Sequencing alleles of the candidates

The presence of polymorphism among the alleles of the candidate gene must be verified before further characterization, not only in the coding region, but also in the regulatory regions. As shown for *Tb1* in maize (see above), the causal polymorphism may reside in the promoter region of the gene. The sequencing of allelic series of the candidate gene in a population with minimum linkage disequilibrium is another means by which to confirm the role of the candidate, by searching for SNPs correlated to the trait. A significant correlation does not imply necessarily a causal relationship, because short-distance linkage disequilibrium may explain the correlation observed, but at least it indicates that the causal polymorphism is close to the SNPs.

Search for QTLs of expression or activity of the candidate

A necessary (but not sufficient) condition to retain a candidate is that its level of expression (mRNA or protein concentration) and/or its activity (for enzyme, transporter, transcription factor, etc.) should be variable in the population studied. In addition, QTLs of expression or activity must be co-located with QTLs of the phenotypic trait under study. The development of high-throughput methods to analyse the transcriptome and proteome expression would make this approach easier.

Transformation

Transformation with DNA fragments carrying the putative QTLs may help to decide between them. However, transformation may pose specific problems with QTLs, because generally the goal is not to complement an inactive gene with an active one, but to compare the effects of two active alleles, which may prove to be statistically difficult. In addition, homologous transformation is not yet possible in plants, so position and dose effects may lead to ambiguous results.

Quantitative complementation

When the QTL of interest has a large effect, and a mutant is known that confers an extreme value on the trait, genetic complementation may be used, by studying a cross between an individual heterozygous at QTL and an individual homozygous for the mutation or a wild/mutant heterozygote. Provided there is dominance of the "high" alleles, segregation in the progeny will be observed only if the QTL and the locus carrying the mutation are the same (respectively 1:1 and 3:1). Otherwise, there will be complementation, the recessive alleles at the two loci being masked, and the population will display no segregation. It is thus that Doebley et al. (1995, 1997) demonstrated that the locus $Tb1$, the mutation of which confers to maize a phenotype similar, in some extent, to that of teosinte, was one of the QTLs responsible for the teosinte-maize differences.

However, this test will be unambiguously performed only in rare cases, because the QTLs have usually weak effects and the dominance, if any, is not complete. So a more general approach, quantitative complementation, may be used, even though the test and its interpretation are more complex (Falconer and MacKay, 1996). Let Q_H and Q_L (H and L for *High* and *Low*) be the two alleles of a QTL detected in a progeny. By marker-assisted backcrossing, a pair of NILs Q_H/Q_H and Q_L/Q_L is developed. On the other hand is developed a pair of NILs wild and mutant for the candidate gene, $+/+$ and m/m, respectively. The crosses $Q_H/Q_H \times +/+$, $Q_L/Q_L \times +/+$, $Q_H/Q_H \times m/m$, and $Q_L/Q_L \times m/m$ will give the genotypes of Table 14, according to whether the candidate corresponds to the QTL (one locus) or not (two loci).

Table 14. Quantitative complementation. NIL_Q and NIL_C correspond to near-isogenic lines for the QTL and for the candidate locus, respectively. Q_X is the allele of the QTL of the lines $+/+$ and m/m, when the candidate gene does not correspond to the QTL.

	The candidate corresponds to the QTL			The candidate does not correspond to the QTL	
	NIL_C $\dfrac{+}{+}$	$\dfrac{m}{m}$		NIL_C $\dfrac{Q_X+}{Q_X+}$	$\dfrac{Q_X m}{Q_X m}$
NIL_Q			NIL_Q		
$\dfrac{Q_H}{Q_H}$	$\dfrac{+}{Q_H}$	$\dfrac{m}{Q_H}$	$\dfrac{Q_H+}{Q_H+}$	$\dfrac{Q_X+}{Q_H+}$	$\dfrac{Q_X m}{Q_H+}$
$\dfrac{Q_L}{Q_L}$	$\dfrac{+}{Q_L}$	$\dfrac{m}{Q_L}$	$\dfrac{Q_L+}{Q_L+}$	$\dfrac{Q_X+}{Q_L+}$	$\dfrac{Q_X m}{Q_L+}$

If the candidate corresponds to the QTL, the difference between the genotypic values of m/Q_H and m/Q_L is expected to be higher than the difference between $+/Q_H$ and $+/Q_L$ (unless allele $+$ is fully recessive towards alleles Q_H and Q_L; if allele $+$ is strictly dominant over Q_H and Q_L, there will be no difference between $+/Q_H$ and $+/Q_L$). If the candidate does not correspond to the QTL, the difference between Q_X+/Q_H+ and Q_X+/Q_L+ is not expected to be significantly different from the one between Q_Xm/Q_H+ and Q_Xm/Q_L+, whatever the dominance between allele Q_X and alleles Q_H and Q_L (the difference is null if Q_X is strictly dominant over Q_H and Q_L). Nevertheless, this test is not unambiguous, because a significant difference may be due to epistatic interaction between the candidate gene and one or several other genes. In that case the candidate would correspond to a QTL affecting the trait, but not the one under study.

CONCLUSION

The first experiments on QTLs in plants date from the mid-1980s. Identification of QTLs will improve our understanding of the molecular and physiological bases of variation of complex traits. In plant breeding, it will enable the monitoring of QTLs themselves, rather than their markers.

The accumulated data on several species reveal clear tendencies, in particular the frequency of cases of apparent co-location between QTLs and genes with qualitative effects (e.g., genes controlling resistance to pathogens, dwarfism, sensitivity to photoperiod, architecture). Even if a huge task remains to be completed case by case to confirm that these involve the same loci, the results, as well as the first examples of QTLs characterized in plants, animals and humans, clearly suggest a continuum between discrete variations and continuous variations. In this context, gene maps are more useful than maps of anonymous markers because they are sources of candidate genes. Programmes of systematic sequencing of EST in plants will help accelerate the development of such maps. In particular, the isolation of cDNA specific to a stage of development, an organ, a stress, and so on, chosen according to the trait studied, can constitute a preliminary test. And again, the recent complete sequencing of *Arabidopsis* and rice genomes represents a huge and invaluable source of information.

5

Molecular Markers in Population Genetics

A. Kremer and S. Mariette

The development of electrophoresis applied to enzymes revolutionized population genetics in the early 1970s. Although the theoretical aspects of this field were already largely developed, notably by the Sewall Wright school (1968, 1969), they could hardly be confirmed with experimental data in the absence of numerous and variable genetic markers. Over the subsequent thirty years, results on allozyme polymorphism accumulated in hundreds, and several bibliographic syntheses were compiled for the plant and animal kingdoms (Nevo et al., 1984; Hamrick and Godt, 1989). The advent of molecular markers considerably accelerated the development of population genetics (Powell, 1994). While allozymes pertain only to a relatively uniform family of markers, access at the DNA level has revealed a considerable heterogeneity of the genome with respect to polymorphism (Chapter 1) and the mode of evolution of different genomic regions. Moreover, polymorphism could be analysed in new compartments (chloroplasts, mitochondria) having their own heredity and evolution. The more traditional field of population genetics, the analysis of the genetic structure of natural populations, was considerably consolidated as a result of information drawn from new molecular markers. The availability of new markers, and particularly the access to DNA sequence data, not only opened up new fields of application, but also made possible the drawing of more refined conclusions on the history and evolution of populations. Phylogenetics is certainly one field that has benefited from the improvement of the DNA marker technology. But the additional comparison of genealogical information with the geographic distribution of species and population (phylogeography) has further permitted the integration of population genetics into a larger historical perspective. It is out of the scope of this work to summarize here the vast applications of molecular markers in population genetics. Rather than making a catalogue of methods used with different markers, we will attempt to use a comparative approach in order to outline their properties as regards the three major applications in population genetics: diversity, differentiation, and gene flow.

SPECIFIC CONTRIBUTIONS OF MOLECULAR MARKERS IN COMPARISON TO TRADITIONAL MARKERS

Studies in population genetics are generally directed towards two major objectives. (1) They can be limited to the descriptive analysis of genetic diversity within a certain number of *genetic units* (e.g., species, populations). (2) Or they further intend to draw evolutionary inferences (population size, migration rates, gene flow, etc.) from the descriptive analysis of diversity. The shift from traditional markers to molecular markers has helped enrich the experiences related to these two objectives, thanks to four qualitative leaps: significant increase in the number of alleles revealed per locus, increase in the number of loci, access to polymorphism in cytoplasmic genomes, and possibility of sequencing and ordering the alleles.

Multiallelism

Enzymatic loci can sometimes have many alleles, with, however, highly unbalanced frequency profiles (Hamrick and Godt, 1989). Recall that although techniques of random amplification (e.g., RAPD, AFLP) reveal only two alleles (for presence or absence of a fragment), other techniques (RFLP, SSCP, DGGE, STS, and SSRs) reveal highly multiallelic loci. In natural populations, 10 to 30 alleles per microsatellite locus can commonly be found (Ashley and Dow, 1994; Rongwen et al., 1995; Dow and Ashley, 1996). It is mostly in the study of the reproductive regime and gene flow that multiallelism of microsatellites proves most useful (Bruford and Wayne, 1993). While allozymes hardly allow us to do more than distinguish self-fertilization from cross-fertilization (Brown et al., 1989), or roughly describe the curves of pollen dispersal (Adams, 1992; Ellstrand, 1992), microsatellites allow us to reconstruct parentage with a very limited number of loci: in a study on oak species, Dow and Ashley (1996) and Streiff et al. (1999) showed that four to six loci were sufficient to identify the parents, even where the parental population was very large. These data can be used for estimation of the size of neighbourhood, the distance over which seeds and pollen are dispersed.

Increase in the number of loci

With isozyme markers polymorphism can be analysed on only a limited number of loci, rarely more than about 30 in plants (Tanksley and Orton, 1983). These limitations are due to technical constraints (enzyme extraction is more delicate in plants because of the presence of numerous secondary metabolites, and above all the number of specific staining reactions is limited) as well as physiological factors, the activity of some enzymatic loci being tissue-specific. The theoretical studies of Nei and Roychoudhury (1974) showed that, for the estimation of diversity, it is necessary to increase the sampling of loci within the genome rather than multiply the number of individuals analysed. Molecular markers, unlike allozymes, allow a more systematic exploration of the genome. For studies of diversity directed towards the biology of conservation, one could estimate the genetic polymorphism within a population more precisely by basing it on a choice of markers taking into account the heterogeneity of the genome. Moreover, the access to a greater number of markers, associated with the availability of genetic maps, ensures better coverage of the genome.

Access to polymorphism of cytoplasmic DNA

Cytoplasmic genomes have certain peculiarities that make them useful in population genetics. Their heredity is generally uniparental. Mitochondria are transmitted preferentially by the female parent. For chloroplasts, the situation is variable. In the Angiosperms, they are maternally inherited in 70% of genera and biparentally in 25% of genera (Harris and Ingram, 1991). In the Gymnosperms, and particularly in the Pinaceae and Taxodiaceae, the mode of heredity varies with the families, but the chloroplasts are preferentially transmitted by the male parent (Hipkins et al., 1994). Even though cytological studies have already suggested differences in parental transmission (Camefort, 1969; Chesnoy, 1987), these facts about the inheritance of chloroplasts were mostly obtained by segregation analysis of cpDNA (chloroplast DNA) or mtDNA (mitochondrial DNA) markers in controlled crosses (Reboud and Zeyl, 1994; Hipkins et al., 1994).

Access to cytoplasmic DNA polymorphism opens a new field of investigation in population genetics (MacCauley, 1995). First of all, differentiation between populations at the nuclear and cytoplasmic levels can be compared to estimate the relative rates of seed flow and pollen flow. Subsequently, the preferentially uniparental inheritance of cytoplasmic organelles simplifies parentage reconstruction. The Pinaceae are an example: the paternal heredity of the chloroplasts and maternal heredity of mitochondria allow us to retrace the dispersal of seeds or pollen by studying only the polymorphism of organelle DNA (Dong and Wagner, 1994). Finally, since plant species can colonize new spaces only by dispersing seeds, the spatial distribution of polymorphism in the genomes of organelles allows us to retrace the colonization routes of the species (Petit et al., 2002). The analysis of polymorphism of chloroplast DNA is facilitated by the advent of "universal" markers appropriate to this genome (see box 5.1).

Ordering of alleles

In the analysis of polymorphism of allozymes, the different alleles are studied without being ordered. The attempts made by Ohta and Kimura (1973) to interpret the electrophoretic variants as alleles derived from one another by shifts of electric charge (stepwise mutation model) proved ineffectual when the model was faced with experimental data, the

---

Box 5.1: "Universal" molecular tools for the study of diversity of chloroplast genomes

The chloroplast genome is remarkably well-conserved within the plant kingdom (Birky, 1988). The organization of genes on the circular molecule of chloroplast DNA (cpDNA) is highly stable from one species to another, and the rate of nucleotide substitutions is lower than in the nuclear and mitochondrial genomes. These characteristics have often contributed to the neglect of cpDNA in the study of diversity within a species. They have, however, enabled the development of "universal" tools in the search for polymorphisms, even rare ones, thanks to the applicability of primers that can be used on a very large number of species, even those that are phylogenetically very distant.

The chloroplast genome has been totally sequenced in several species (gymnosperms, dicotyledons, monocotyledons), which has enabled the discovery of highly conserved genomic regions. The detailed study of such regions, by sequence alignment, has led to the identification of primers in the conserved regions, from which polymorphic fragments can be amplified (see Chapter 1, subsection *A particular case of RFLP: CAPS*). The products of amplification are then digested by restriction enzymes to reveal the polymorphism. More than 38 pairs of primers have thus been developed that cover the entire large single copy region of the chloroplast genome (Demesure et al., 1995; Dumolin-Lapegue et al., 1996; Grivet et al., 2002). They have been tested on 20 plant species belonging to 13 families.

A similar approach was adopted to discover microsatellite motifs in the gymnosperms (Powell et al., 1995). By analysis of the complete sequence of the chloroplast genome of *Pinus thunbergii*, 20 pairs of primers flanking microsatellite motifs were constructed (Vendramin et al., 1996). These 20 pairs proved effective for all the other pines tested, and for other genera of the family Pinaceae: 65% of primers for the genus *Abies*, 75% for the genera *Cedrus* and *Picea*. Conserved microsatellite motifs have also been discovered in angiosperms (Weising and Gardner, 1999).

---

differences in charge not being solely responsible for the electrophoretic mobility (Fuerst and Ferrell, 1980). In the case of molecular markers, relations between alleles can be established as a function of the mode of mutation. For tandem repeat regions (mini- and microsatellites, ribosomal DNA), the alleles can be derived from one another by addition or subtraction of one or several repeat units (Shriver et al., 1993), and they can thus be ordered according to their size (stepwise mutation model). Nevertheless, taking into account this particular mode of evolution, two alleles of the same length are not necessarily identical by descent. For other genomic regions, each mutation can reasonably be considered unique (model of infinite number of alleles). Two alleles of identical sequence are thus identical by descent. In this case, a phylogenetic tree of alleles can be constructed from their sequences. Unless there has been little recombination (cpDNA, for example), one can also construct a phylogenetic tree from maps or restriction profiles of different alleles.

In addition to the use of ordering of alleles to refine the calculation of genetic diversity, the availability of DNA sequences of the different alleles allows us to tackle the genetics of populations in a "retrospective" way by looking at the genealogy of alleles (theory of coalescence, Hudson, 1990). Many data can be deduced from these sequences, notably the time of coalescence (number of generations separating the alleles from their common ancestor) and the positioning of different nodes on the genealogical tree of alleles. The coalescence approach offers further a probabilistic method to infer demographic and genetic parameters from gene genealogies (Slatkin, 1991).

ANALYSIS OF MOLECULAR DIVERSITY

The great variety of markers available has led to several methods of analysis of diversity in natural populations. These methods are divided into two major groups. The first is limited to the *phenotypic* interpretation of data. The differences between genetic units are quantified by indexes of dissimilarity (Lynch, 1990), and multivariable statistical methods allow us to describe the general organization of the diversity. This approach is used when markers resulting from genetic fingerprinting techniques are available. The nature of markers justifies only qualitative comparisons between different studies: since the loci are not sought to be individualized, their scope is limited to the case studied. In the second group, the data are analysed in *genetic* terms, using codominant markers (e.g., RFLP, STS, SSRs, SNPs) or dominant markers (e.g., RAPD, AFLP) by treating the loci revealed *individually*. In the rest of this chapter the genetic interpretation will be detailed, and not the phenotypic analysis. For reasons already mentioned, the genetic analysis is richer and notably allows for inferences on the evolution and structure of populations. From the analysis of nucleotide polymorphism, molecular diversity can be quantified, by way of the average number of substitutions per nucleotide site. This number can be estimated from sequencing data or from data resulting from molecular markers.

Measurement of nucleotide divergence from sequences

Consider two alleles, of known sequence, which present a difference at sites 5 and 14:

Site	5	14
Allele 1	ACCTGCTATCTTACGACGGTCGCGATGATA	
Allele 2	ACCTCCTATCTTAGGACGGTCGCGATGATA	

Suppose that the two alleles evolved separately over time T (T can also be the number of generations). Each sequence will undergo mutations, which will result in the replacement of certain nucleotides by others. Let K be the average number of substitutions per nucleotide site that are produced over time T, n the size of sequences, and n_d the number of nucleotidic sites that show differences between the two sequences after time T ($n_d = 2$ when alleles 1 and 2 are compared in the example above). If the rate of nucleotide substitution is low, then the number of substitutions observed after time T will be close to that really produced, i.e., $K = n_d/n$ (2/30 in our example). If, on the other hand, the rate of substitution is high, certain substitutions may be produced recurrently at the same nucleotide sites. For example, even if the alleles 1 and 2 have the same base A at site 8, it is possible that several substitutions have occurred during time T, but finally the result is the same. The average number of substitutions observed after time T will be less than that really produced. Jukes and Cantor (1969), Kimura (1980), and Tajima and Nei (1984) proposed methods to estimate K taking into account the recurrent substitutions. These methods have comparable results when $K < 0.5$, and the simplest method (Jukes and Cantor, 1969) is the most commonly used:

$$K = \left(-\frac{3}{4}\right) \ln \left(1 - \frac{4}{3} \frac{n_d}{n}\right) \qquad (1)$$

At lower values of K (< 0.10), the direct method can be used ($K = n_d/n$) (Nei, 1987). K is also called the *nucleotide divergence* between the two sequences.

The measurement of K has great evolutionary significance. If λ is the rate of nucleotide substitution per unit of time (equal to rates of mutation m in the framework of the neutralist theory), we have:

$$K = 2\lambda T \qquad (2)$$

(The factor 2 comes from the fact that the substitutions are produced on one or the other of the sequences.) The concept of "molecular clock" originates in this linear relation between the number of accumulated substitutions and time. It is used to estimate the time of divergence between sequences (Table 15).

Measurement of nucleotide divergence using marker techniques

Marker techniques derived from the use of restriction enzymes or PCR amplifications have in common that they reveal a sequence polymorphism located in a recognition site depending on the technique used. It is a restriction site in the case of RFLP, a primer hybridization site for RAPD, and the two together for AFLP (Chapter 1). Through the set of recognition sites, molecular markers thus allow measurement of nucleotide diversity

Table 15. Comparison of nucleotide divergence for the gene coding for the large unit of ribulose 1,5-bisphosphate carboxylase (*rbcL*) between different species of Angiosperms and one Gymnosperm (*Pseudotsuga menziesii*) (Bousquet *et al.*, 1992).

Species	Size of the sequence	Number of observed substitutions	$K1$	$K2$	$K3$
Zea mays (A)	1422	232	0.163	0.184	0.186
Triticum aestivum (A)	1425	227	0.159	0.179	0.181
Serenoa repens (P)	985	122	0.124	0.135	0.137
Magnolia macrophylla (P)	1425	165	0.116	0.126	0.128
Spinacia oleracea (A)	1425	214	0.150	0.168	0.171
Quercus rubra (P)	1425	173	0.121	0.132	0.134
Itea virginica (P)	1407	166	0.118	0.128	0.129
Petunia hybrida (A)	1425	209	0.147	0.163	0.166

A, annual plant. P, perennial plant. $K1$, number of substitutions per nucleotide sequence ($K = n_d/n$, see text). $K2$, number of substitutions per site with the correction of Jukes and Cantor (1969) (see equation 1 in text). $K3$, number of substitutions per site with the correction of Kimura (1980). The corrections, which take into account recurrent substitutions at the same nucleotide site, are justified by the fact that the values of $K1$ are relatively high (see text). It is noted that the two methods of correction give very similar results. The nucleotide divergences vary according to the perennial or annual character of the species studied. The molecular evolution of the gene coding for *rbcL* will be more rapid in annual plants than in perennial plants.

on the scale of a genome. The estimation method is based on the hypothesis that any sequence substitution on a recognition site (or between two recognition sites) is expressed by the presence or absence of a band on an electrophoretic profile. The average proportion of bands shared by two sequences, G, must thus in principle be an indicator of the average number of nucleotide substitutions on the recognition sites. The details of the method are given by Nei and Li (1979), Nei and Miller (1990), and Gonzales-Candelas et al. (1995) for the RFLP, by Clark and Lanigan (1993) for RAPD products, and Innan et al. (1999) for AFLP. Only the principles are summarized here.

Consider a recognition site of given length r (r = the number of bases). If the time of divergence T between two sequences A and B is not too long, then the number of nucleotide substitutions at a given site follows a Poisson distribution of expectation λT. The probability P that the recognition site remains unchanged during the course of period T is:

$$P = e^{-r\lambda T} \tag{3}$$

Using equation (2), we obtain for K:

$$K = \left(\frac{-2}{r}\right) \ln P \tag{4}$$

Nei and Li (1979) moreover showed that a simple relation can be established between P and G:

$$P \approx G(3 - 2P_0)^{1/4} \tag{5}$$

P can be found from an iterative method taking $(P_0 = G^{1/4})$ for the initial value. G can be directly estimated from the electrophoretic profiles by:

$$G = 2M_{AB} / (M_A + M_B) \tag{6}$$

where M_{AB} designates the number of bands common to the two sequences A and B, and M_A and M_B designate the number of bands of sequences A and B, respectively.

It is considered that for a band to subsist over the evolutionary process, the recognition sites must remain unchanged and at the same time there must not be insertion-deletion polymorphism. In the case of RFLP analysis, the comparison between the direct method (equation 2) and the indirect method (equation 5) on simulated sequences showed that the second was applicable for values of K less than 0.05, and that the use of restriction enzymes with 4 bases was preferable to that of enzymes with 6 bases (Kaplan, 1983; Gonzalez-Candelas et al., 1995).

The use of the method for amplification products, especially RAPD, seems more difficult, even if the results of simulation are encouraging (Clark and Lanigan, 1993). Indeed, the homology of amplified fragments is not always certain, and a nucleotide substitution (notably in the 5' region of the site) does not necessarily prevent amplification. Estimations of nucleotide divergence may therefore be underestimated. This risk is undoubtedly smaller in the case of AFLP, since the additional nucleotides are at the 3' end, where mismatching rules out amplification. In the case of AFLP, theoretical predictions indicated that the method for estimating nucleotide diversity at the whole genome level based on AFLP electrophoretic profiles is reliable when K is less than 0.10 (Innan et al., 1999); however, the variances of the estimation of diversity increase as K increases.

Finally, almost all of these approaches were developed in the case of haploid genomes. In order to extend them to diploid genomes, as considered by Clark and Lanigan (1993), the allelic frequencies must be known, which is more difficult in the case of dominant markers (see next section).

The main interest of the indirect method is that it offers the possibility of comparing genetic parameters estimated with different markers at the whole genome level. As we will see in the following sections, the parameters of diversity can be defined at the nucleotide level, in parallel with what has been done at the allelic level. The "reduction" of these parameters to the nucleotide level makes comparisons possible between different markers.

POLYMORPHISM WITHIN A POPULATION

Estimation of the divergence between two alleles mentioned in the previous section is extended in this section to the set of alleles present in a population. Intrapopulation polymorphism results as a balance of different evolutionary forces: mutation at the source of polymorphism, migration tending to increase it, and genetic drift tending to diminish it. We have seen that different techniques are not equally effective in revealing polymorphism. The level of polymorphism observed will thus depend on the rates of mutation in the region considered as well as on the marker technique used. We present here the definitions of parameters of polymorphism and refer to other authors (Brown and Weir, 1983; Weir, 1990; Kremer et al., 1997; for their statistical properties.

Expression of polymorphism within a population

Polymorphism is generated by the presence of multiple variants within a genetic unit. The variants are, according to the techniques used, haplotypes or nucleotides, while the genetic units are genotypes, populations, or species. The notion of polymorphism is thus relative to a marker and to a genetic unit. The indexes of polymorphism used in the literature, in ecology (Pielou, 1975) as well as in population genetics (Nei, 1987; Brown and Weir, 1983), are grouped in two major categories: those that take into account the number of variants within the genetic units analysed, and those that are based on the relative frequency of variants.

INDEXES TAKING INTO ACCOUNT THE NUMBER OF VARIANTS

The number of variants is extremely sensitive to sample size (Gregorius, 1980) and cannot be used in cases where sizes vary significantly, even if statistical techniques are used to adjust that index to constant sizes (Hurlbert, 1971; El Mousadik and Petit, 1996a). The problems are still more acute when the markers used are characterized by the presence of a highly frequent variant and many rare variants (case of isozymes). Moreover, the number of variants can be uninformative, particularly when it is bounded to an upper limit (2 in the case of RAPD or AFLP).

INDEXES TAKING INTO ACCOUNT THE FREQUENCY OF VARIANTS

The index most frequently used is derived from that introduced by Simpson (1949) in ecology. It was introduced in population genetics by Marshall and Jain (1969) and generalized under the term *genetic diversity* (*h*) by Nei (1973, 1987). The genetic diversity is defined as the probability with which two variants drawn at random within a genetic unit will be different. This index is popular in population genetics because it is equivalent to the rates of heterozygosity in a population in panmictic equilibrium, and because its value depends essentially on the most frequent alleles, thus limiting the samples required to estimate it.

Other indexes have been proposed: e.g., effective number of alleles (Crow and Kimura, 1964), information function of Shannon-Weaver (Lewontin, 1972), and proportion of heterozygous loci. They are all derived, in one way or another, either from the number of variants or from the genetic diversity. Moreover, the knowledge they provide is limited to the description and comparison of values obtained in different genetic units. Few inferences, in terms of evolution of populations, can be made from these indexes. In contrast, the fluctuation in number of variants and diversity during evolution of populations has been widely studied (Nei, 1987). For populations evolving under the model of Wright-Fisher (constant population size, absence of migration between populations, model of mutation corresponding to an infinite number of alleles with non-ordered alleles), equilibrium values of number of alleles (A) and genetic diversity (h) are written respectively as follows (Nei, 1987; Hartl and Clark, 1989):

$$A = \sum_{k=1}^{k=x} \frac{4N\mu}{4N\mu + k - 1} \tag{7}$$

$$h = 4N\mu / (4N\mu + 1) \tag{8}$$

where N is the effective size of each of the populations, μ is the rate of mutation, and x is the sample size.

In the absence of migration, the level of diversity thus depends on the drift and rates of mutation. As drift acts in the same way for all the markers, the differences in diversity observed between markers are thus essentially due to differences in rates of mutation (Tables 16 and 17). Finally, exhaustive bibliographic syntheses of values obtained for these two indexes, compiled from allozymic data, have set them up as standards of measurement of diversity (Hamrick and Godt, 1989).

Measurement of genetic diversity with different types of markers

AT THE ALLELIC LEVEL

The most common expression of genetic diversity is that given by Nei (1973). If the variants considered are haplotypes or alleles at a given locus with A variants, then this expression is:

$$h = 1 - \sum_{i=1}^{A} p_i^2 \tag{9}$$

Table 16. Measurement of the level of intrapopulation diversity using different types of markers

Species	Isozymes			RFLP			RAPD			References
	nl	A	H	nl	A	H	nl	A	H	
Brassica campestris	3	2.7	0.518	9	8.5	0.684				McGrath and Quiros, 1992
Brassica oleracea	4		0.318				64		0.488	Lanner-Herrera et al., 1996
Hordeum vulgare	7	4.86	0.439	13	4.46	0.469				Zhang et al., 1993
Pinus sylvestris	20		0.269				22		0.327	Szmidt et al., 1996
Picea mariana	13	2.2	0.300				28	1.9	0.321	Isabel et al., 1995
Populus grendidata	14	1.4	0.080	37	1.8	0.13	56	1.9	0.310	Liu et Fournier, 1993
Populus tremuloides	13	2.8	0.250	41	2.7	0.250	61	2	0.300	Liu et Fournier, 1993
Quercus petraea	8	3.3	0.381				31	1.9	0.233	Lecorre et al., 1997
Zea mays	15	1.69	0.179	35	3.46	0.472				Dubreuil and Charcosset, 1998

nl, number of loci analysed. *A*, average number of alleles per locus. *h*, genetic diversity within a population. These comparative studies were done on the same sample (populations and individuals) and with the same estimation methods (equation 9). The differences observed for *A* and *h* cannot thus be attributed to different samples. On the other hand, the intragenome sampling (number of loci) is highly unbalanced between the markers, with notably an insufficient number of enzymatic loci. As a general rule, the genetic diversity (*h*) is higher when it is estimated from molecular markers than from isozymes. The rate of mutation of the former is undoubtedly greater than that of the latter. This comes from the fact that non-coding regions undergo fewer selective constraints than coding regions. And if the markers are in the genes, the degeneration of the genetic code, and the fact that all the substitutions of amino acids are not detectable, explain this difference. It is not the same for allelic richness (RFLP-isozyme comparison), even though in general the number of alleles seems higher for the RFLP (nevertheless, the comparison is difficult, since the number of alleles revealed in RFLP is directly linked to the number of restriction enzymes used).

Table 17. Correlation coefficient between diversity assessed with two different markers on the same set of populations

Reference	Species	N_P	N	Marker 1	n_1	Marker 2	n_2	Correlation	Probability
Dubreuil and Charcosset, 1998	Zea mays	10	30	Isozyme	20	RFLP	35	0.466	0.125
McGrath and Quiros, 1992	Brassica campestris	20	?	Isozyme	5	RFLP	4	0.544	0.025
Pogson et al., 1995	Gadus morhua	6	81 to 138	Isozyme	10	RFLP	17	0.630	0.176
Zhang et al., 1993	Hordeum vulgare	19	?	Isozyme	7	RFLP	18	0.562	0.050
Barker et al., 1997 (a and b)	Bubalus bubalis	11	20 to 57	Isozyme	53	SSR	21	0.667	0.083
Cagigas et al., 1999	Salmo trutta	4	20 to 25	Isozyme	10	SSR	3	0.400	0.603
Raybould et al., 1999	Brassica oleracea	5	20 to 100	Isozyme	4	SSR	7	0.900	0.089
Scribner et al., 1994	Bufo bufo	3	10	Isozyme	6	SSR	1	-0.500	0.289
Sun et al., 1998	Elymus fibrosus	10	3 to 5	Isozyme	12	SSR	6	0.159	0.660
Scribner et al., 1994	Bufo bufo	3	10	Isozyme	6	Minisatellite	3	0.500	0.724
Desplanque et al., 1999	Beta vulgaris	6	5	RFLP	4	SSR	1	0.928	0.044
Scribner et al., 1994	Bufo bufo	3	10	SSR	1	Minisatellite	3	-1.000	0.077
Aagaard et al., 1998	Pseudotsuga menziesii	6	13 to 36	Isozyme	20	RAPD (G1, F=0)	36	0.493	0.300
Baruffi et al., 1995	Ceratitis capitata	6	?	Isozyme	26	RAPD (?)	176	0.418	0.191
Cagigas et al., 1999	Salmo trutta	4	20 to 25	Isozyme	10	RAPD (G1, F=0)	46	0.800	0.225
Cagigas et al., 1999	Salmo trutta	4	20 to 25	Isozyme	10	RAPD (G2, F=0)	19	-0.800	0.119
Díaz et al., 2000	Elymus fibrosus	10	3 to 5	Isozyme	12	RAPD (G1, F=0)	60	0.045	0.877
Ferguson et al., 1998	Lens spp.	5	100	Isozyme	8	RAPD (G1, F=0)	14	0.359	0.537
Huang et al., 1998	Castanea dentata	5	15	Isozyme	13	RAPD (G1, F=0)	31	0.200	0.863
Isabel et al., 1995	Picea mariana	12	11 to 30	Isozyme	14	RAPD (codominant)	22	0.400	0.484
Lannér-Herrera et al., 1996	Brassica oleracea	18	38 to 60	Isozyme	4	RAPD (P)	64	-0.100	0.764
Le Corre et al., 1997	Quercus petraea	21	23	Isozyme	8	RAPD (G2, F isozymes)	31	0.392	0.080
Nadler et al., 1995	Ascaris suum	7	?	Isozyme	3	RAPD (G1, F=0)	18	0.214	0.630
Papa et al., 1998	Hordeum vulgare	12	20	Isozyme	6	RAPD (?)	77	0.685	0.024
Wu et al., 1999	Pinus attenuata	4	5 to 50	Isozyme	36	RAPD (G1, F=0)	98	0.200	0.863
Wu et al., 1999	Pinus muricata	5	10 to 36	Isozyme	32	RAPD (G1, F=0)	98	0.684	0.205
Wu et al., 1999	Pinus radiata	3	15 to 28	Isozyme	32	RAPD (G1, F=0)	91	0.866	0.384
Cagigas et al., 1999	Salmo trutta	4	20 to 25	SSR	3	RAPD (G1, F=0)	46	0.800	0.225

Reference	Species	N_p	N	Marker 1	n_1	Marker 2	n_2		
Cagigas et al., 1999	*Salmo trutta*	4	20 to 25	SSR	3	RAPD (G2, F=0)	19	-0.200	0.603
Sun et al., 1998	*Elymus fibrosus*	10	3 to 5	SSR	6	RAPD (G1, F=0)	60	0.679	0.031
Thomas et al., 1999	*Pinus contorta*	9	40	SSR	5	RAPD (G1, F=0)	10	0.500	0.368
Yan et al., 1999	*Aedes aegypti*	3	40	RFLP	8	AFLP (G1, F=0)	137	1	0.290
Cagigas et al., 1999	*Salmo trutta*	4	20 to 25	RAPD (G1, F=0)	10	RAPD (G2, F=0)	19	-0.400	0.387
Mariette et al., 2001	*Pinus pinaster*	23	30	SSRs	3	AFLP (G2, F=SSRs)	122	-0.272	0.341
Mariette et al., 2002b	*Quercus petraea*	7	50 to 200	SSRs	6	AFLP (G2, F=SSRs)	155	-0.649	0.103
Mariette et al., 2002b	*Quercus robur*	7	50 to 200	SSRs	6	AFLP (G2, F=SSRs)	155	0.180	0.691

N_p is the number of populations, N is the mean sample size of the populations, n_1 and n_2 are the number of loci studied for marker 1 and 2. For dominant markers (AFLP, RAPD), data were analysed either by method G1 or G2. G1 corresponds to the calculation of diversity estimates following the estimation of allelic frequencies; G2 is the same analysis but discarding loci for which the frequency of homozygotes for the null allele is less than $3/N$ (recommendation of Lynch and Milligan, 1994). F is the fixation index used to infer allelic frequencies for dominant markers. Genetic diversities are only poorly correlated when estimated with different marker techniques: only in 4 out of 36 case studies were the correlations found significant (Type I error: 0.05). Low correlation coefficients are most likely due to poor sampling efforts, especially for the number of loci. Other causes can be related to low differentiation between populations for their diversities (Mariette et al., 2002a).

which is also written as

$$h = \sum_i^A \sum_{j \neq i}^A p_i p_j$$

where p_i (or p_j) is the frequency of allele i (or j).

In this equation, the alleles are not ordered in any way. If they can be ordered as a function of distance Δ_{ij} between alleles i and j on the basis of external information, the expression of the diversity becomes:

$$h = \sum_i^A \sum_{j \neq i}^A p_i p_j \Delta_{ij} \qquad (10)$$

It is important to note that this quantity is no longer the probability with which two alleles drawn at random from a population will be different, but an average distance between alleles present in a population.

For the majority of markers, there is no information that will enable the ordering of alleles. In this case Δ_{ij} equals 1 for all pairs of alleles considered, and expression (10) becomes equivalent to expression (9). On the other hand, for markers comprising tandem repeats (e.g., microsatellites, rDNA), the shift from one allele to another often occurs by the addition or suppression of a repeat unit; hence, the number of repeat units separating two alleles may constitute a measure of their distance. Finally, when DNA sequences of alleles are available, Δ_{ij} can be derived from their nucleotide divergence K (equations 1 and 4). It is up to the user to judge the opportunity for introducing the distance between alleles in the measurement of the diversity, depending on the purpose of the study. When Δ_{ij} is taken into account, a consensus must be defined for the unit of measure chosen if different studies are to be compared.

The particular case of dominant markers
To measure allelic diversity for dominant markers (RAPD, AFLP), the frequencies of alleles 1 (presence of fragment) and 0 must be known. In the case where the population is in Hardy-Weinberg equilibrium, the frequency q of the null allele (absence of fragment) can be deduced from the frequency Q of individuals in which the electrophoretic profiles do not show this fragment: $q = \sqrt{Q}$. In the case where the population is not in Hardy-Weinberg equilibrium, the frequency q becomes (Chong et al., 1994; Le Corre et al., 1997):

$$q = \left(\frac{-1}{2(1-F)} \right) \left(F - \sqrt{F^2 + 4(1-F)Q} \right) \qquad (11)$$

where F is the index of fixation, which measures the deviation between the theoretical heterozygosity at equilibrium, He, and the observed heterozygosity, Ho ($F = 1 - Ho/He$) (Hartl and Clark, 1989). An estimation of F can be obtained from codominant markers available in the same species. Obviously, equation (11) can only be used if the index of fixation F is the same for the codominant markers used to estimate F and that for which q is calculated. Strictly, equation (11) can thus be applied only for neutral markers, and at comparable rates of mutation. If no information on F is available in the same population, or if there are doubts about the validity of values obtained from other markers, then it is always possible to estimate F from values of Q estimated on two successive generations of the same population (Lynch and Milligan, 1994).

The estimation of the frequency of the null allele from frequencies of homozygotes for the null allele shows, however, a significant bias when the number of homozygote genotypes is less than 3 (Lynch and Milligan, 1994). These authors recommend that loci for which the frequency of homozygotes for the null allele is less than $3/N_0$ should not be retained, N_0 being the sample size, and that the sample sizes be increased 2 to 10 times in comparison to codominant markers. These recommendations also apply to the estimation of diversity. The use of dominant markers to assess genetic diversity requires the sampling of more loci than with codominant markers, and the elimination of the least polymorphic loci. The increase in the number of loci that enables multilocus marker techniques such as RAPD or AFLP could compensate for the loss in precision at each locus. This has recently been investigated by simulation studies (Mariette et al., 2002a). It is shown that despite the poor precision at a single locus basis, AFLPs are preferable in estimating diversity at the whole genome level, because the random distribution accounts for the enormous heterogeneity of diversity existing within the genome. A compilation of comparative studies of diversity conducted with different types of markers indicates that the diversity estimates were significantly correlated in less than 12% of the cases (Table 17). Simulation studies conducted by Mariette et al. (2002a) suggested that the lack of correlation may be due to (1) a low sample size of loci used, (2) a low differentiation of populations for their level of diversity, or (3) non-equilibrium situations.

At the nucleotide level

The principle of estimation of diversity at the nucleotide level develops directly from that used at the allele level. Take the example of alleles of known sequences and constant length n, and consider each nucleotide site as a locus with four "alleles" (the four bases A, T, G, C).

At a site k, the nucleotide diversity can be calculated in the following manner (Nei and Miller, 1990; Tajima, 1993):

$$h_k = 1 - \sum_{i=1}^{4} x_{ik}^2 \tag{12}$$

where x_{ik} is the frequency of nucleotide i (i varying from 1 to 4 for the bases A, T, G, and C respectively) at site k. The average nucleotide diversity over n sites constituting the sequence is:

$$h = \frac{1}{n} \sum_{k}^{n} h_k \tag{13}$$

This expression is equivalent to that using the frequencies of different haplotypes p_i and p_j:

$$h = \sum_{i} \sum_{j \neq i}^{n} p_i p_j \, \pi_{ij} \tag{14}$$

where π_{ij} is the nucleotide divergence between the haplotypes i and j (K of equation 1). Nei and Miller (1990) recommend that π_{ij}, the mean number of nucleotide substitutions, be estimated by $\pi_{ij} = n_d/n$ when π_{ij} is less than 0.02, and by the approximation of Jukes and Cantor (equation 1) when π_{ij} is greater than 0.02.

Comparison of equations (12) and (14) with equation (10) shows that the *diversity at the nucleotide level* is equivalent to the *diversity at allelic level when the alleles are ordered as a function of their nucleotide divergence*.

DIFFERENTIATION BETWEEN POPULATIONS

Diversity is generally analysed over several populations rather than a single one. Two hierarchic levels can thus be considered: the diversity between populations, generally called *differentiation*, and intrapopulation diversity, which we have mentioned in the preceding section. Generalization is possible at other hierarchic levels. The degree of differentiation between populations depends very closely on the age of the separation of populations and the gene exchanges maintained between them. Thus, knowledge of the degree of differentiation provides information about the history of populations.

General expression of differentiation

The differentiation can be expressed in two ways, which yield the same result (Nei, 1987). The first consists of considering the distances d_{xy} between all the pairs of populations x and y:

$$d_{xy} = h_{xy} - \frac{h_x + h_y}{2} \tag{15}$$

where h_x (or h_y) is the diversity in population x (or y) (equations 9, 10, 13, or 14 according to the level considered), and h_{xy} is the interpopulation diversity (when alleles i and j in the equations cited are drawn from two different populations x and y). From this, the global differentiation is deduced (d_{st}) as the average of all the terms d_{xy} out of all the pairs of populations (including reciprocal couples and those associating a population with itself):

$$d_{st} = \frac{1}{s^2} \sum_x^s \sum_y^s d_{xy} \tag{16}$$

where s is the number of populations considered.

The second approach consists in partitioning the genetic diversity at two hierarchic levels (individual populations and bulk of populations). The diversity over the bulk of populations considered as a single population (called total diversity or h_t) is calculated by taking for frequencies in equations (9), (10), (13), or (14) the mean frequencies over all the populations. The average diversity of individual populations is calculated as the average of intrapopulation diversities (h_s). It can be shown (Nei, 1987) that the differentiation between populations (equation 16) is expressed also as the difference of diversities estimated at two hierarchic levels.

$$d_{st} = h_t - h_s \tag{17}$$

The differentiation G_{st} is then expressed relative to the total diversity:

$$G_{st} = \frac{d_{st}}{h_t} \tag{18}$$

With a simplified model of evolution of populations (without selection), where populations of the same size all exchange an equal number of genes (island model of Wright, 1969), the value expected at equilibrium between migration and drift is written as follows (Nei, 1987):

$$G_{st} = \frac{1}{1 + \left(4N\left(\frac{s}{s-1} \right)(\mu + m) \right)} \tag{19}$$

where m is the migration rate. If the mutation rate is much lower than the migration rate, and if the number of populations sampled is large, this relation takes the more familiar form (Hartl and Clark, 1989):

$$G_{st} \approx \frac{1}{1 + 4Nm} \tag{20}$$

The parameter of differentiation G_{st} is defined in the framework of a model with fixed effects: only the differentiation within the sampled populations is considered. The model with random effects was introduced by Weir and Cockerham (1984; see also Weir, 1990) in the form of parameter θ, expressed at the level of an allele (θ_W at the level of a locus), equivalent to F_{st} of Wright (1951), coefficient of correlation between genes of different individuals of a single population. The two models differ in their underlying hypotheses, the definition of parameters of differentiation, and the methods of estimation. While Weir and Cockerham made the hypothesis that the populations all derive from a single ancestral population and that they have evolved without migration, mutation, or selection, Nei makes no hypothesis on the history of the populations. Finally, Pons and Petit (1995), in the case of haploid data, and Pons and Chaouche (1996), in the case of diploid data, extend the decomposition of Nei (equation 17) to the random effect model, without the highly constraining genetic hypotheses of the Weir and Cockerham model. It is thus a model reconciling the two preceding approaches. The comparisons made from experimental data show that the estimates of the two parameters (G_{st} and F_{st}) converge when the number of populations is high (Chakraborty and Leimar, 1987; Pons and Petit, 1995). In the literature, the parameters of differentiation at the allelic level and at the nucleotide level are inspired by one or the other of these three reference models, which are hereafter designated NE for Nei, WE for Weir and Cockerham, and PO for Pons and Petit.

Measurement of differentiation with different marker systems

Equations (15) and (18) constitute a general framework for expressing the differentiation that can be applied to all the markers, at the allele level as well as the nucleotide level.

At the allele level

The values of G_{st} are obtained by calculating h_t and d_{st} using equations (9) and (16) or (17), with different weights according to the model used. As with diversity, the distance between alleles can be introduced (equation 10), as was done in the framework of the WE model (Excoffier et al., 1992) and the PO model (Pons and Petit, 1995). In the particular case of microsatellites, if the stepwise mutation model is accepted, then the differentiation can be calculated by taking the difference in the number of repeat units as a distance between alleles. The estimation methods presented in the literature and applied to the particular case of microsatellites are derived from the NE model (parameter R_{st}, Slatkin, 1995) or the WE model (parameter Φ_{st}, Michalakis and Excoffier, 1996).

At the nucleotide level

The differentiation at the nucleotide level is obtained by replacing the terms in h in equation (15) by those given in equations (12) or (14). This is the expression given by Nei (1982) under the term γ_{st}, which corresponds to the NE model. Weir and Basten (1990) used analysis of variance to estimate the differentiation at the nucleotide level (parameter β analogous to parameter θ of the WE model). The other methods available are inspired more by the PO model. Lynch and Crease (1990) and Hudson et al. (1992), excluding the comparisons of populations with themselves in equation (14), approach a random effects model (N_{st} for the parameter defined by the former, $< F_{st} >$ for the parameter of the latter). These two parameters differ from each other only in whether or not they take into account the correction of Jukes and Cantor (equation 1). Holsinger and Mason-Gamer (1996) proposed an estimator without the bias of Nei (1982), which renders γ_{st} comparable to the F_{st} of Wright (1951).

In the same manner as for diversity, the *differentiation at the nucleotide level* is equivalent to the *differentiation at the allelic level when the alleles are organized as a function of their nucleotide divergence*.

The experimental results published so far chiefly pertain to sequence data (Mason-Gamer et al., 1995; Byrne and Moran, 1994). Nevertheless, the method of estimation of nucleotide divergence from data resulting from restriction fragments or random amplifications (equation 4) can be used to estimate the nucleotide differentiation, provided that the hypotheses indicated are respected.

In theory, the differentiation is relatively independent of the type of marker used, if the markers are neutral. Its value at equilibrium is not affected by the mutation rate when that rate is negligible compared to the rate of migration (equation 20). Hence, differentiation depends mainly on the population size and rate of migration, which constitute factors acting in the same sense and with the same amplitude on the different markers. Experimental results comparing the values of differentiation obtained with different markers tend to confirm this observation (Table 18). However, for highly polymorphic markers (microsatellites), lower differentiation than with other markers is expected, because the rate of mutation is very high and can certainly no longer be discarded. Theoretical studies conducted with the help of simulations indicate, moreover, that G_{st} is independent of the number of alleles at the marker analysed, reinforcing thus the "universal" nature of this parameter (McDonald, 1994).

GENE FLOW

With traditional genetic markers as isozymes, analysis of gene flow was limited to indirect estimates of the number of migrants as inferred from

Table 18. Population differentiation assessed with two different markers on the same set of populations.

Reference	Species	N_P	N	Marker 1	n_1	Marker 2	n_2	Differentiation 1	Differentiation 2
Dubreuil and Charcosset, 1998	Zea mays	10	30	Isozyme	20	RFLP	35	0.230	0.220
McGrath and Quiros, 1992	Brassica campestris	20	?	Isozyme	5	RFLP	4	0.448	0.403
Pogson et al., 1995	Gadus morhua	6	81 to 138	Isozyme	10	RFLP	17	0.014	0.069
Zhang et al., 1993	Hordeum vulgare	19	?	Isozyme	7	RFLP	18	0.296	0.476
Barker et al., 1997(a and b)	Bubalus bubalis	11	20 to 57	Isozyme	53	SSR	21	0.157	0.168
Cagigas et al., 1999	Salmo trutta	4	20 to 25	Isozyme	10	SSR	3	0.472	0.328
Estoup et al., 1998	Salmo trutta	11	25 to 30	Isozyme	8	SSR	3	0.258	0.279
Raybould et al., 1999	Brassica oleracea	5	20 to 100	Isozyme	4	SSR	7	0.072	0.234
Ross et al., 1997	Solenopsis invicta	2	70 to 284	Isozyme	11	SSR	7	0.117	0.038
Scribner et al., 1994	Bufo bufo	3	10	Isozyme	6	SSR	1	0.090	0.016
Sun et al., 1998	Elymus fibrosus	10	3 to 5	Isozyme	12	SSR	6	0.650	0.540
Tessier et al., 1995	Salmo salar	2	30	Isozyme	1	SSR	4	0.099	0.006 to 0.341
Scribner et al., 1994	Bufo bufo	3	10	Isozyme	6	Minisatellite	3	0.090	0.015
Huang et al., 2000	Haliotis rubra	9	10	Isozyme	3	Minisatellite	2	0.078	0.016
Scribner et al., 1994	Bufo bufo	3	10	SSR	1	Minisatellite	3	0.016	0.015
Aagaard et al., 1998	Pseudotsuga menziesii	6	13 to 36	Isozyme	20	RAPD (G1, F=0)	36	0.290	0.340
Ayres et Ryan, 1997	Wyethia reticulata	4	3 to 17	Isozyme	22	RAPD (G1, F=0)	21	0.190	0.250
Ayres et Ryan, 1999	Wyethia reticulata	4	10 to 20	Isozyme	8	RAPD (G1, F=0)	21	0.250	0.250
Ayres et Ryan, 1999	Wyethia bolanderi	4	10 to 20	Isozyme	8	RAPD (G1, F=0)	20	0.070	0.120
Bucci et al., 1997	Pinus leucodermis	7	17 to 30	Isozyme	16	RAPD (G1, F=0)	23	0.067	0.179
Buso et al., 1998	Oryza glumaepatula	4	50	Isozyme	4	RAPD (P)	15	0.310	0.640
Buso et al., 1998	Oryza glumaepatula	4	50	Isozyme	4	RAPD (G2, F isozymes)	13	0.310	0.670
Cagigas et al., 1999	Salmo trutta	4	20 to 25	Isozyme	10	RAPD (G2, F=0)	19	0.472	0.237
De Wolf et al., 1998	Littorina striata	5 to 6	57 to 73	Isozyme	4 to 5	RAPD (G1, F=0)	6	0.001 to 0.032	0.002 to 0.076
Díaz et al., 2000	Elymus fibrosus	10	3 to 5	Isozyme	12	RAPD (G1, F=0)	60	0.700	0.630
Ferguson et al., 1998	Lens culinaris	5	20	Isozyme	8	RAPD (G1, F=0)	27	0.800	0.780

Reference	Species	N_P	N	Marker 2	n_2	Marker 1	n_1		
Ferguson et al., 1998	Lens odemensis	5	20	Isozyme	8	RAPD (G1, F=0)	26	0.950	0.995
Ferguson et al., 1998	Lens nigricans	5	20	Isozyme	8	RAPD (G1, F=0)	14	0.920	0.790
Ferguson et al., 1998	Lens ervoides	5	20	Isozyme	8	RAPD (G1, F=0)	15	0.950	0.950
Ferguson et al., 1998	Lens lamottei	5	20	Isozyme	8	RAPD (G1, F=0)	20	0.150	0.820
Isabel et al., 1995	Picea mariana	5	15	Isozyme	13	RAPD (codominant)	31	0.033	0.053
Lanner-Herrera et al., 1996	Brassica oleracea	18	38 to 60	Isozyme	4	RAPD (P)	64	0.370	0.140
Le Corre et al., 1997	Quercus petraea	21	23	Isozyme	8	RAPD (G2, F_{IS} isozymes)	31	0.027	0.024
Mamuris et al., 1999	Mullus surmuletus	7	40 to 45	Isozyme	3	RAPD (?)	18	0.035	0.053
Nadler et al., 1995	Ascaris suum	6	40 to 49	Isozyme	20	RAPD (G1, F=0)	15	0.094	0.092
Papa et al., 1998	Hordeum vulgare	12	20	Isozyme	6	RAPD (?)	77	0.161	0.107
Vicario et al., 1995	Abies alba	7	27 to 35	Isozyme	11	RAPD (?)	55	0.160	0.110
Wu et al., 1999	Pinus attenuata	4	5 to 50	Isozyme	36	RAPD (G1, F=0)	98	0.220	0.360
Wu et al., 1999	Pinus attenuata	4	5 to 50	Isozyme	36	RAPD (G2, F=0)	27	0.220	0.170
Wu et al., 1999	Pinus muricata	5	10 to 36	Isozyme	32	RAPD (G1, F=0)	98	0.320	0.450
Wu et al., 1999	Pinus muricata	5	10 to 36	Isozyme	32	RAPD (G2, F=0)	43	0.320	0.270
Wu et al., 1999	Pinus radiata	3	15 to 28	Isozyme	32	RAPD (G1, F=0)	91	0.120	0.260
Wu et al., 1999	Pinus radiata	3	15 to 28	Isozyme	32	RAPD (G2, F=0)	31	0.120	0.170
Yan et al., 1999	Aedes aegypti	3	40	RFLP	8	AFLP (G1, F=0)	13	0.063	0.033
Cagigas et al., 1999	Salmo trutta	4	20 to 25	SSR	3	RAPD (G2, F=0)	19	0.328	0.237
Huang et al., 2000	Haliotis rubra	9	10	SSR	3	RAPD (P)	84	0.078	0.086
Sun et al., 1998	Elymus fibrosus	10	3 to 5	SSR	6	RAPD (G1, F=0)	60	0.540	0.630
Thomas et al., 1999	Pinus contorta	12	40	SSR	5	RAPD (G1, F=0)	10	0.028	0.061
Mariette et al., 2001	Pinus pinaster	23	30	SSR	3	AFLP (G2, F=SSRs)	12	0.111	0.061
Mariette et al., 2002b	Quercus petraea	7	50 to 200	SSR	6	AFLP (G2, F=SSRs)	15	0.023	0.044
Mariette et al., 2002b	Quercus robur	7	50 to 200	SSR	6	AFLP (G2, F=SSRs)	15	0.020	0.030
Huang et al., 2000	Haliotis rubra	9	10	Minisatellite	2	RAPD (P)	84	0.086	0.016

N_P is the number of populations, N is the sample size in the populations, n_1 is the number of loci studied for marker 1, and n_2 is the number of loci studied for marker 2. For dominant markers (AFLP, RAPD), data were analysed either by method G1 or G2. G1 corresponds to the calculation of diversity estimates following the estimation of allelic frequencies; G2 is the same analysis but discarding loci for which the frequency of homozygotes for the null allele is less than $3/N$ (recommendation of Lynch and Milligan, 1994). F is the fixation index used to infer allelic frequencies for dominant markers. In most case studies there is a high congruence of differentiation estimated with different markers, because differentiation is rather independent of marker-specific properties (mutation rates or others, see text).

differentiation measures among neighbouring populations. These estimates were extremely crude as they depended heavily on an unrealistic evolutionary scenario (most generally the island model) and as they were cumulative over successive generations. These indirect methods were improved by subdividing the total gene flow into pollen and seed migration rates thanks to the separate estimation of nuclear and cytoplasmic DNA differentiation. In addition, the development of highly polymorphic codominant markers has now made it possible to reconstruct parentage over two successive generations and consequently retrace pollen and/or seed movement.

Indirect measures of gene flow

COMPARATIVE RATES OF SEED AND POLLEN MIGRATION

The uniparental heredity of cytoplasmic genes results in the decrease of the effective size of populations to at least half in comparison to nuclear genes (Birky, 1988, 1991). The decrease, associated with weaker gene flow between populations, is expressed by a greater sensitivity to genetic drift and must thus induce a higher differentiation between populations for the cytoplasmic genes. These intuitive predictions have been confirmed by analytic developments (Birky et al., 1989; Petit et al., 1993a). Figure 43 shows the theoretical comparative evolution of the differentiation for chloroplast markers and nuclear markers. Three important results can be drawn from these curves:

- The differentiation of a nuclear marker is clearly smaller than the differentiation of a chloroplast marker when the transmission of chloroplasts is preferentially maternal ($\alpha > 0.5$, where α is the rate of transmission by the female parent).
- When there is equal transmission by the two parents ($\alpha = 0.5$), the differentiation of a chloroplast marker is of the same order of magnitude as that of a nuclear marker.
- The differentiation of a chloroplast marker can increase abruptly, as soon as even a minimal transmission is produced by the other parent ($\alpha = 0.95$ as opposed to $\alpha = 1$).

These theoretical developments allow us to link the ratio of rates of migration by seeds and by pollen with equilibrium G_{st} values (island model for the cytoplasmic and nuclear genes) for a marker with maternal heredity and a marker with biparental heredity with (Ennos, 1994; El Mousadik and Petit, 1996b):

$$\frac{m_p}{m_g} = \frac{2(1Gst_m - 1) - (1/Gst_n - 1)}{(1 - 1/Gst_m)} \tag{21}$$

where m_p and m_g denote respectively the rates of migration by pollen and

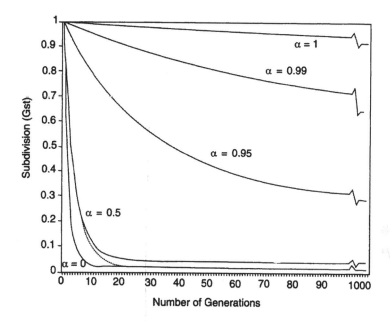

Figure 43. Variations in G_{st} for a strictly allogamous plant in the theoretical case of a set of populations sharing an equal number of migrant genes (island model) but asymmetrically between seeds and pollen (ratio of migration rates ≈ 500 in favour of pollen). The rates of transmission by the female parent (α) have also been taken into account (α = 1 corresponds to an exclusively maternal transmission). The populations are supposed to be totally differentiated at the start (G_{st} = 1) and of a constant size of 100 individuals. Dotted line: differentiation for a nuclear marker. Solid lines: differentiation for a chloroplast marker (Petit et al., 1993a).

by seeds, and Gst_m and Gst_n are the differentiation values for a marker with maternal and biparental heredity, respectively. Table 19 illustrates the contrast between differentiation at the nuclear level and differentiation of a genome inherited maternally, as well as the asymmetry of gene flows by these two vectors.

ORIGIN OF COLONIZING GENOTYPES

The analysis of differentiation in subsets of populations undergoing different colonization dynamics has been shown to be helpful in identifying the origin of colonizing genotypes. The evolution of the differentiation was described in the theoretical case of a group of populations exchanging an equal number of migrants among all the populations (island model of Wright, 1969). In this model, at the equilibrium between migration and drift, the differentiation is of the same order of magnitude whatever the subset of populations considered. Foundation is yet more pronounced at the margin of the distribution area, where the species is in the process of colonization, and thus must result in higher levels of differentiation than at the centre of the distribution area. This observation brought Whitlock and McCauley (1990) to express analytically the differentiation between recently formed populations (Fst_o in the expression of McCauley):

$$Fst_o = (1/2k) + \Phi(1 - 1/2k)Fst_c \tag{22}$$

where k denotes the number of colonizers that have formed the new population, Φ the probability that two genes drawn at random among the founders come from the same source population, and Fst_c the differentiation between populations already established (the source populations).

If Φ tends to zero, the differentiation between the populations founded is chiefly generated by the number of founders, and colonization tends to destroy the structure that existed beforehand. The founders consist of individuals originating from different populations. In contrast, if Φ tends to 1, the founders have issued from a limited number of populations and colonization tends to increase the differentiation. The knowledge of Φ allows us thus to obtain valuable information on the dynamics of colonization.

As shown by McCauley et al. (1995), k can be measured by demographic data, Fst_o and Fst_c can be calculated from a sample of populations recently founded and a sample of established populations, and Φ can be deduced from equation (22). In an example given by McCauley et al. (1995), the values of Φ were 0.73 for allozymes and 0.89 for chloroplast DNA, indicating that the colonizers come preferentially from a single source population. The difference between the two values of Φ suggests, moreover, that several colonizers originated from the same parents in the source population.

Direct measures of gene flow

In contrast to the indirect measures of gene flow inferred from differentiation among populations, instantaneous measures are independent on evolutionary models and refer to migration occurring between the last two generations. With the steady increase of the development of microsatellites, genetic fingerprints could be assigned to each genotype of a given natural population. Fingerprints can then be compared among individuals and genetic relationships can be looked for when the fingerprints are compatible. Parentage analysis simply consists then in assigning parent-offspring relationships between members of a juvenile and adult cohort in a natural population. When the markers are highly variable, the assignment can be made after excluding all non-compatible pairs of genotypes. This procedure results most generally in assigning as parent the adult genotype that exhibits a compatible multilocus genotypic array with a juvenile plant. If simple exclusion methods cannot be used, e.g. when more than one adult genotype is compatible, maximum likelihood methods are used and the most probable parents are selected according to the likelihood scores (Meagher and Thompson, 1986). Beforehand, one has generally delineated an area in which all adult plants are genotyped and to which genotypes of the juvenile cohort are compared. If the studied area is in a continuous population, it may happen that no compatibility at all is found, and the real parent is then considered to be outside the study area. In addition, if the marker system is not polymorphic enough, it may also happen that the compatible parent genotype identified inside the study area is actually the wrong parent and that the real parent is outside the area (cryptic gene flow). Computer packages are now available that allow us to calculate likelihood ratios for the different putative parents, and to estimate cryptic gene flow and exclusion probabilities (Marshall et al., 1998; Gerber et al., 2000). In plants, parentage analysis can face practical limitations. In the case of temperate trees, it was found that paternity analysis resulted in the identification of less than 50% of male parents on a studied area of 5 to 10 ha where all adult genotypes were scored. Laboratory resources do not usually allow the scoring of more than some hundreds of adult genotypes, therefore reducing the size of the studied area. Hence, for species that exhibit long distance pollen flow, the method ignores more than half of the pollination events (Dow and Ashley, 1998; Streiff et al., 1999). Second, when parentage analysis is performed, e.g., paternity and maternity analyses, the method allows identification of the pair of parents, but not separation of the male and the female parent. Cytoplasmic markers, if polymorphic within the area, may be useful in this case. An alternative method is to analyse separately the DNA of the seed coat and the embryo, since the seed coat has the diploid genotype of the maternal plant (Godoy and Jordano, 2001).

As parentage analysis may require extensive genotyping, alternative methods have been searched for recently that would avoid the genotyping of all putative parents. The proposal is to compare the allele frequencies in pollen pools pollinating different female plants sampled in a population. The male gametic heterogeneity among female parents, expressed as a genetic differentiation coefficient Φ_{ft}, has been found to be inversely proportionate to the average pollen dispersal and to the distance separating the mothers (Smouse et al., 2001; Austerlitz and Smouse, 2001). When the distance between maternal plants is large enough (higher than 5 times the average distance of pollen dispersal), then Φ_{ft} becomes independent of the distance separating the mother plants and is only related to the pollen dispersal distance. The method does not require highly polymorphic markers, but if male and female multilocus arrays cannot be unambiguously separated in offspring then the method can be biased. More recently, the method has been refined and Φ_{ft} can be computed between pairs of maternal parents, improving the method especially for markers with low exclusion probabilities.

CONCLUSION

The use of different molecular markers to describe genetic polymorphism

Table 19. Nuclear differentiation (Gst_n), chloroplast differentiation (Gst_m), and ratio of rates of migration by pollen (m_p) and by seed (m_g) in different species (data provided by R.J. Petit)

Species	Gst_n	Reference	Gst_n	Reference	m_p/m_g
Argania spinosa	0.25	El Mousadik and Petit, 1996a	0.60	El Mousadik and Petit, 1996a	2.5
Eucalyptus nitens	0.30	Moran, 1992	0.62	Byrne and Moran, 1994	1.8
Hordeum spontaneum	0.48	Brown and Munday, 1982	0.75	Neale et al., 1988	1.3
Quercus petraea	0.024	Zanetto et al., 1994	0.90	Petit et al., 1993	500
Quercus robur	0.032	Zanetto et al., 1994	0.92	Petit et al., 1993	286
Silene alba	0.16	McCauley et al., 1995	0.63	McCauley et al., 1995	7

In conformity with theoretical predictions, the differentiation for a nuclear gene, transmitted by the two parents, is always less than the differentiation for a cytoplasmic gene transmitted by the seeds. This difference accounts for differences in rate of migration by the pollen vector and by the seed vector. The asymmetry between these rates also seems to conform to that suggested by the biology of species. Species with heavy seeds (oaks) disseminate their seeds over short distances (dispersal by gravity, even though a small number of seeds may be dispersed to longer distances by birds such as jays). Thus, there is greater asymmetry in such species.

in natural populations limits comparisons across species and prevents any general conclusion on the structure of genetic diversity. Meta-analyses were performed within a given family of markers (Hamrick and Godt, 1989, for isozymes; Nybom and Bartish, 2000, for RAPD). Interestingly, they revealed similar trends of variation between species for diversity and differentiation when compared with life history traits of the species studied. However, trends of variation within species can be different when diversity surveys are conducted with different markers on the same set of populations. As shown in Table 17, levels of diversity were only rarely significantly correlated. The low correlations observed between different markers undoubtedly underline the deficiencies in sampling loci and the very high heterogeneity of polymorphism within the genome. Efforts should be made to monitor the genetic diversity at more loci. As expected in theory, in contrast to diversity, differentiation among populations exhibits rather congruent results when different types of markers are used. Markers have contributed greatly to the study of gene flow in plants, thanks to the comparative analysis of polymorphism in the nuclear and cytoplasmic compartments. There as well, theoretical predictions of the evolution of cytoplasmic genomes have been confirmed by experimental results: the nuclear differentiation is always less than the cytoplasmic differentiation (Table 19). The rapid development of microsatellite markers (in nuclear genomes and even in chloroplast genomes, see box 5.1) will make it possible to estimate effective distances of migration of pollen and seeds, and the reproductive success of members of a population.

6

Application of Markers in Selection

A. Charcosset and A. Gallais

One of the fields concerned with molecular markers is plant selection. Two major types of application can be identified. First, markers can provide a new picture of the structure of diversity within a species or a species complex (Chapter 5). This information is particularly important for genetic resource management, as well as for the rational use of genetic resources in selection programmes. The most promising applications of this type are outlined in the first part of this chapter. Second, markers make it possible to construct dense genetic maps that allow us to locate loci having a qualitative effect (Chapter 3) or a quantitative effect (Chapter 4), as well as to estimate the effects of such loci. This information can have direct repercussions on selection, the simplest of which is the use of markers of genes to guide the introgression of alleles of interest in recipient genomes. This point is addressed in the second part of the chapter.

CONTRIBUTION OF DIVERSITY STUDIES TO SELECTION

Some of the methods described in Chapter 5 can be applied to the analysis of the genetic diversity of collections of genetic resources (further referred to as accessions, generally inbred lines or populations) and have an immediate value in selection. We will describe first the relationships existing between divergence at the level of markers, phenotypic divergence, and coefficient of coancestry. From these relationships, a number of applications can be proposed. We will then address the question of application of markers in the framework of two objectives related to the creation of hybrid varieties: classification of the material into heterotic groups and prediction of the hybrid value. Finally, the use of these techniques with respect to plant variety protection will be outlined.

Principal methods of data analysis

Among the methods for studying diversity, calculation of genetic distance has been the object of particular interest for plant breeders. When data are interpreted from allelic variation at given loci, various indexes of distance can be estimated from allelic frequencies (de Vienne and Damerval, 1985). The distance between two populations most commonly used is (Rogers, 1972):

$$D_{xy} = \frac{1}{2L} \sum_{l=1}^{L} \sum_{a_l=1}^{n_l} (f_x^{a_l} - f_y^{a_l})^2 \tag{1}$$

where L is the number of loci used to calculate the distance, n_l is the total number of alleles at locus l, and $f_x^{a_l}$ (or $f_y^{a_l}$) is the frequency of the allele a_l at locus l in population x (or y). This expression is equivalent to equation (15) in Chapter 5. In case of pure lines, D_{xy} is simply equal to the proportion of loci for which the lines have fixed different alleles. When the information cannot be analysed in terms of allelic variation (for example when it is coded in presence-absence of bands), the distance of Nei and Li (1979) (equations 4 to 6 in Chapter 5) is conventionally used. When only part of the information can be analysed in terms of allelic variation, it may be advantageous to calculate a weighted average of distances of Rogers and Nei and Li (Dillmann et al., 1997a). The precision of distance estimation increases with the number of loci that are considered (Table 20). It has to be noted that confidence intervals are not symmetrical (unless the estimated distance is 50%). For example, with 10 markers, a real distance (over the entire genome) of 30.8% can lead to a zero value of the estimated distance with a probability of 2.5% $[(1 - 0.308)^{10}]$, so that the confidence interval for a nil distance is $[0\% - 30.8\%]$. For a given number of markers, an even distribution of the markers on the genetic map also generally improves the precision of distance estimation, when compared to a random sampling of markers (Dillmann et al., 1997b; Wang and Bernardo, 2000). It also has to be noted that the map distance between genetic markers can be taken into

Table 20. Precision of genetic distance estimation between two lines as a function of the number (N) of loci used (Bar-Hen and Charcosset, 1994).

N	$\hat{D}$	0%	10%	20%	30%	50%
10		0 – 30.8	0.2 – 44.5	2.5 – 55.6	6.6 – 65.2	18.8 – 81.2
20		0 – 16.8	1.2 – 31.7	5.7 – 43.6	11.8 – 54.3	27.2 – 72.8
30		0 – 11.5	2.1 – 26.5	7.7 – 38.5	14.7 – 49.4	31.3 – 68.7
50		0 – 7.1	3.3 – 21.8	10.0 – 33.7	17.8 – 44.5	35.6 – 64.4
100		0 – 3.6	4.9 – 17.6	12.7 – 29.2	21.2 – 40.0	39.9 – 60.1
200		0 – 1.8	6.2 – 15.0	14.7 – 26.2	23.7 – 36.9	42.9 – 57.1

account to increase the precision of genetic distance estimation, using appropriate statistical tools (Dillmann et al., 1997b).

The genetic distance can serve as a basis for the classification of the materials of interest   (e.g., lines, populations). Many algorithms are presently used, the description and comparison of which would be beyond the scope of this survey. The most commonly used are those of the average linkage (commonly called UPGMA), where the distance between two groups is defined as the mean distance between elements of the groups, and the method of Ward (1963), which minimizes the variation within the groups formed. These analyses can be complemented by other descriptive approaches of the variability, such as principal component analysis or factorial analysis of correspondences.

Relationships between molecular divergence, phenotypic divergence and coancestry

A strong relationship between distance and coancestry is generally observed when a sufficiently large number of genetic markers is considered (see Melchinger, 1998 for a review). Low distances can be observed only for high levels of coancestry (see, for example, Smith et al., 1990). Nevertheless, for a similar degree of coancestry, the divergence between two individuals is variable because of genetic drift and effect of selection. For example, when recombinant inbred lines are generated by successive selfing from a cross between two lines A and B, some lines are closer to A than others, even if all have the same coancestry coefficient with A. This can be illustrated by results from Bernardo and Kahler (2001) showing that, in maize, contribution of a given parent to inbreds obtained in such a way may vary between 80% and 20%. Relative to the information about coancestry, the genetic markers have the advantage of taking into account these effects of drift and selection and therefore provide a more accurate evaluation of genetic similarity.

Many studies have attempted to state precisely the relationship between indexes of distance estimated at the phenotypic level and indexes of distance estimated from genetic markers. These studies first used isozyme markers (Damerval and de Vienne, 1985), then molecular markers (Bar-Hen and Charcosset, 1994; Dillmann et al., 1997a; Burstin et al., 1995; Burstin and Charcosset, 1997; Moser and Lee, 1994). Generally, the results show "triangular" relationships between the two types of distances (Fig. 44): while two individuals similar at the level of markers almost always have similar phenotypes, a large distance at the level of markers does not allow for any inference as to the resemblance or phenotypic divergence. This relationship can be explained by the polygenic bases of the traits generally taken into account in the estimation of the phenotypic distance. If one considers four biallelic loci with equal effects, and if the favourable and unfavourable alleles are designated + and − respectively, then the lines

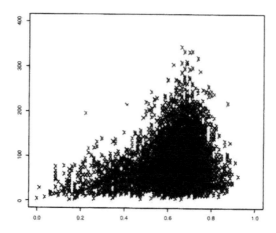

Figure 44. Relationship between distance estimated from RFLP marker loci (X-axis) and morphological distance (Y-axis) for a collection of maize lines (Dillmann et al., 1997a).

carrying combinations of alleles ++ − − and − − ++ will have the same phenotype, even if they differ at each locus (see Burstin and Charcosset, 1997, for a detailed analysis of this relationship).

These properties of the relationship between indexes of distance estimated at the phenotypic level and indexes of distance estimated from genetic markers have two significant practical consequences. First, the agronomic traits of an individual can be predicted if it is similar, with respect to genetic markers, to an individual of known phenotype. This is particularly interesting for phenotypic traits that are costly or time-consuming to evaluate. This prediction can be made empirically or formalized through appropriate statistical methods (see next section for heterosis prediction). The second application lies in the possibility of distinguishing whether two individuals or accessions that are close phenotypically are similar at the level of genetic markers. If it is assumed that marker loci provide a representative sample of the entire genome, individuals with a high similarity at the level of markers are likely to carry similar alleles for most QTL, whereas individuals with a high divergence are more likely to carry different alleles. This is particularly important for the conservation and use of genetic resources. Accessions that are similar at both the phenotypic and marker levels can be considered redundant so that some of them can be either eliminated from the collection or bulked to facilitate the conservation. On the other hand, accessions that are similar at the phenotypic level but distant at the marker can be considered original.

Application of classification methods: heterotic groups and genetic resource management

The first experimental studies considering the creation of hybrid varieties in maize indicated a relationship between the performance of the hybrids and the origin of their parental lines (initial populations or area of culture). First, it was observed that crosses between lines that are related (by pedigree) generally led to less productive hybrids than those obtained from non-related lines (Hayes and Johnson, 1939; Krug et al., 1943). Moreover, some pairs of populations proved complementary with respect to hybrid breeding, when lines developed from one population generally resulted in good hybrids when crossed with lines of the second population (Krug et al., 1943). On the basis of these empirical results, breeders defined heterotic groups, including lines with comparable crossing behaviour and often with a common origin (for example, a given traditional open-pollinated population). In this way, the groups Lancaster and Reid Yellow Dent, largely used in the Corn Belt, were defined in the United States.

Some authors have studied representative sets of lines from different heterotic groups by means of genetic markers, the statistical analysis of results having been conducted with the help of the methods described earlier: hierarchical classification from molecular distances or factorial analyses. In all the studies conducted on maize, the groups established from the molecular information were consistent with classifications into heterotic groups (Godshalk et al., 1990; Dudley et al., 1991; Melchinger et al., 1991, 1992; Boppenmeier et al., 1992; Livini et al., 1992; Messmer et al., 1992; Dubreuil et al., 1996). Results of the same type were observed in sunflower (Gentzbittel et al., 1994; Tersac et al., 1994;). For example, Figure 45 illustrates the divergence between the American groups and the European flint lines in maize. This differentiation between groups reflects their divergence at the level of allelic frequencies for a series of loci. The results have two major consequences for selection. If well-established heterotic groups have been defined on the basis of a large number of empirical studies, genetic markers can be used to determine the position of a genetic material of uncertain origin, with respect to those groups. It seems that about 60 multiallelic loci should be sufficient to answer the question (Dubreuil et al., 1996). This assignation to groups facilitates the choice of tester(s) to be crossed with this material for the purpose of estimating its combining ability. If there are no established heterotic groups within a species of interest, as is the case with newly selected species or species for which hybrid development is recent, an analysis by genetic markers can reveal a pre-existing structure (see, for example, the results of Bark and Havey, 1995, on onion). As a second step, it is necessary to evaluate which pairs of groups are the most promising for the creation of high-performing hybrids.

Even though the objective is not to define heterotic groups, analysis of the structure of the genetic diversity and definition of groups within a species can prove to be very useful in rationalizing the management and exploitation of genetic resources in selection (Keim et al., 1989; Deu et al., 1994; Melchinger et al., 1994). The classifications can be the basis from

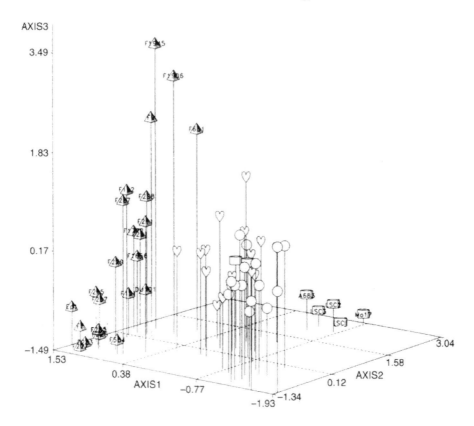

Figure 45. Associations revealed within a maize inbred lines collection by principal component analysis of RFLP data. The symbols identify different heterotic groups (pyramids for flint lines of European origin, other symbols for the heterotic American groups) (after Dubreuil et al., 1996).

which "pools" can be constituted by intercrossing units of a single group. Such pools can be used to initiate "prebreeding" programmes of improvement designed to enlarge the variability used in creating varieties (Gallais, 1990). The classifications can also be used as a basis to define "core collections", aiming to represent the diversity of the entire collection from a limited number of accessions (Brown, 1989). This can be achieved, for example, by the selection of a constant number of accessions within each group formed by the classification process. Other approaches have been proposed to optimize sampling through iterative processes (van Hintum, 1994; Bataillon et al., 1996; Gouesnard et al., 2001). Such core collections are helpful in genetic resource management, facilitate exchanges between programmes or countries, and can be used to identify the units from a larger collection that present particular characteristics (e.g., resistance to a disease).

Relationship between molecular divergence and heterosis

It is possible to show theoretically (Falconer, 1961) that the heterosis of a hybrid, defined as the difference between the performance of that hybrid and the average performance of its parents, is related to the divergence of the parents for QTLs displaying dominance effects. In the framework of a biallelic model, the heterosis H_{xy} observed for a cross between two populations x and y is written as follows:

$$H_{xy} = \sum_{l=1}^{L} \sum_{a_l=1}^{2} \frac{1}{2} d_l (f_x^{a_l} - f_y^{a_l})^2 \qquad (2)$$

where d_l is the effect of dominance at QTL l (the other variables are the same as in equation (1)). In F_1 hybrids, the heterosis is linked to the heterozygosity at the QTLs that show dominance effects. If the loci have equal dominance effects, the heterosis will be directly proportionate to the mean heterozygosity. On the basis of this theoretical analysis, as well as the empirical results mentioned above on the effect of inbreeding, many geneticists have attempted to predict the heterosis from indexes of divergence between parents. The results obtained from morphological criteria have been generally disappointing (see Lefort-Buson, 1985b). This could be explained partly by the reasons mentioned earlier: lines having the same value for a quantitative trait may have different alleles for each of the underlying loci (Charcosset, 1992). In this case, the heterosis predicted for the hybrid between two lines of this type will be nil while the heterosis really observed will be maximal.

We have seen that genetic markers, unlike quantitative traits, carry information on the divergence of genotypes at individual loci. The mean heterozygosity of a hybrid can thus be estimated directly from the analysis of its two parental lines. This property has motivated several studies attempting to predict the heterosis through indexes of distances estimated using genetic markers (Lefort-Buson, 1985a). A synthesis of results obtained in maize and in other species is presented in Table 21. Several trends can be pointed out:

- Overall, the predictions made using numerous RFLP markers seem more effective than those made from a limited number of enzymatic markers (see particularly the study of Smith et al., 1990), which can be explained by the better "coverage" of the genome and a higher average number of alleles per locus for RFLP.
- When significant correlation coefficients are observed, indexes of divergence show a greater relationship with the specific combining ability (Sprague and Tatum, 1942, also called specific heterosis by Gardner and Eberhart, 1966) than with the performance of hybrids. This result can be explained with the help of simple models of quantitative genetics. The hybrid value depends partly on QTL having

Table 21. Correlation coefficients observed between the distances estimated using isozyme markers (Iso), RFLP, RAPD, STS, AFLP or SSR and: (1) hybrid value (HV); (2) specific combining ability (SCA); and (3) mid-parent heterosis (HE). The number of loci considered for distance estimation is indicated within brackets following the technique considered. * and ** denote correlation coefficients significant at 5% and 1% respectively.

Species	Trait	Parameter	Markers	Correlation	References
Maize	Grain yield	HV	Iso (11)	0.09	Hunter and Kannenberg, 1971
Maize	Grain yield	SCA	Iso (8)	0.23	Heidrich-Sobrinho and Cordeiro, 1975
Maize	Grain yield	HV	Iso (13)	0.14	Price et al., 1986
Maize	Grain yield	HV	Iso (11)	0.20**	Lamkey et al., 1987
Maize	Grain yield	HV	Iso (13)	0.60**	Smith and Smith, 1989
"	"	HV	RFLP (157)	0.93**	"
Maize	Grain yield	HV	RFLP (33)	0.46*	Lee et al., 1989
"	"	SCA	RFLP (33)	0.73*	"
Maize	Grain yield	HV	RFLP (82)	0.18	Melchinger et al., 1990a
"	"	SCA	RFLP (82)	0.26	"
Maize	Grain yield	HV	RFLP (82)	0.32**	Melchinger et al., 1990b
"	"	SCA	RFLP (82)	0.39**	"
Maize	Grain yield	HV	RFLP (47)	0.09	Godshalk et al., 1990
Maize	Grain yield	HV	RFLP (66)	0.10	Dudley et al., 1991
Maize	Grain yield	SCA	RFLP (66)	0.35*	"
Maize	Grain yield	HV	RFLP (149)	0.36**	Ajmone-Marsan et al., 1998
"	"	SCA	RFLP (149)	0.65**	"
"	"	HV	AFLP (209)	0.51**	"
"	"	SCA	AFLP (209)	0.72**	"
Rice	Grain yield	HV	Iso (6)	−0.26	Peng et al., 1988
Rice	Grain yield	HV	RFLP (68) + SSR (8)	0.48**	Zhang et al., 1994
"	"	HE	RFLP (68) + SSR (8)	0.77**	"
Rice	Grain yield	HV	RFLP + SSR (105)	0.31	Zhang et al., 1995
Rice	Grain yield	HE	RFLP + SSR (105)	0.53	"
Rice	Potential yield	HV	RAPD (74) + SSR (22)	0.78**	Xiao et al., 1996a
Rice	Grain yield	HV	RFLP + SSR (108)	0.79**	Saghai Maroof et al., 1997
"	"	HE	RFLP + SSR (108)	0.47*	"
Wheat	Grain yield	HV	STS (27)	0.27	Martin et al., 1995
Oat	Grain yield	HE	RFLP (26)	−0.30; 0.08	Moser and Lee, 1994
Brassica juncea	Grain yield	HV	Iso (9)	0.29*	Sekhon and Gupta, 1995
Brassica juncea	Grain yield	HV	RAPD (34)	ns	Jain et al., 1994
Rapeseed	Grain yield	HV	RFLP (38)	0.49*	Diers et al., 1996
"	"	HE	RFLP (38)	0.58*	"
Soybean	Grain yield	HV	RFLP (129)	−0.29	Cerna et al., 1997
"	"	HE	RFLP (129)	0.08	"
Sunflower	Grain yield	HV	AFLP (360)	0.63	Cheres et al., 2000

uniquely an additive effect, for which heterozygosity does not confer any advantage, while the specific heterosis does not depend on these loci. This result is interesting to the extent that the specific combining ability is the component of the hybrid value that is the most difficult to predict from "classical" agronomic designs. General combining ability of a given inbred can indeed be estimated rather accurately by the average performance of its hybrid progenies with a set of relevant testers. The results are closely dependent on the material studied. First of all, the highest correlation values have been obtained for studies including consanguineous hybrids (crosses between related lines). The study of Smith and Smith (1989) includes hybrids having a high coefficient of consanguinity (around 0.9) and leads to a strong correlation. On the other hand, the study of Godshalk et al. (1990) does not include consanguineous hybrids and does not result in a significant relationship. The correlation diminishes when consanguineous hybrids are withdrawn from the set of hybrids studied (Charcosset, 1992; Burstin et al., 1995; Ajmone-Marsan et al., 1998). These results confirm a trend indicated at the isoenzyme level by Frei et al. (1986) in maize and subsequently by Sekhon and Gupta (1995) in mustard. Finally, for studies done on heterotic "intergroup" hybrids (see above), there does not appear to be any relationship between the value of the hybrid and the divergence of the parents (Melchinger et al., 1992; Boppenmeier et al., 1992). The same result is observed at the intersubspecific level in rice (Xiao et al., 1996a; Sagai Maroof et al., 1997) (Fig. 46).

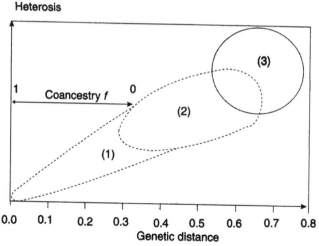

Figure 46. Schematic representation between mid-parent heterosis for yield and parental genetic distance based on unselected DNA markers covering the entire genome: (1) crosses between related lines; (2) intra-group crosses between unrelated lines; (3) inter-group crosses (Melchinger, 1999) Genetic Diversity and Heterosis.

This variety of results can be analysed theoretically. First of all, the genetic markers used are generally considered phenotypically neutral. For them to have a predictive value, there must be linkage disequilibrium between the marker alleles and the alleles of QTL (Charcosset et al., 1991). The disequilibrium is particularly high within populations that have been developed recently from a limited number of parents. This can be illustrated by the results of Stuber et al. (1992), who developed an F_3 mapping population from the cross between lines B73 and Mo17. Each F_3 plant studied was crossed with both parental lines, and the average productivity of the corresponding hybrid progenies was evaluated. That productivity was highly correlated (correlation of 0.68) to the average heterozygosity at marker loci. Within a genetic group, linkage disequilibrium may exist if the group has been created recently from a limited number of parents or has undergone significant drift phenomena. In allogamous species, assuming a panmictic type of reproduction, the linkage disequilibrium will decrease over generations, leading to a reduction in the relationship between distance and heterosis (Charcosset et al., 1991).

The case of hybrids between lines originating from different genetic groups is particularly interesting. For a relationship to exist between divergence and performance for these hybrids, not only must there be linkage disequilibrium between markers and QTL within each of the groups considered, but also the associations between marker alleles and QTL alleles must be the same in the different groups (Charcosset and Essioux, 1994). The lines of different groups having generally been derived from different populations, and having evolved independently for a very long time, this second condition is not likely to be fulfilled in most cases. Finally, when intra- and intergroup hybrids are considered simultaneously, the differentiation between the genetic groups may explain the existence of linkage disequilibrium (the specific alleles of groups are linked among themselves), even if the disequilibrium is nil within each of the groups. The possibility of being able to distinguish the intragroup hybrids from intergroup hybrids thus gives some efficiency to the prediction, as can be illustrated by the results obtained by Nienhuis and Sills (1992) in bean.

Other genetic factors may act on the intensity of the relationship between heterosis and molecular divergence, notably epistasis and the relative importance of effects of dominance at QTL. However, a combination of experimental and theoretical results concur to propose a general view of the relationship between heterosis and molecular divergence and the conditions under which prediction is expected to be efficient (Fig. 46).

Finally, the importance of hybrids between unrelated inbred lines has led to the search for other approaches to the prediction of heterosis. These approaches are based on the idea developed earlier that if two hybrids A × B and A′ × B′ have parents closely related at the level of genetic markers (A′ close to A and B′ close to B), their heterosis or specific combining abilities will be similar. Bernardo (1994) developed a method based on BLUP (best linear unbiased prediction, Henderson, 1975) to formalize this approach. The distances between lines are here used to estimate the coefficients of coancestry (Bernardo, 1993), which are then used to estimate covariances between individuals whose performance is known, on the one hand, and individuals whose performance needs to be predicted, on the other hand. Then, from these covariances and the known performances, predictions can be made as desired using an appropriate formula. Experimental results demonstrate that this approach can be efficient in situations where the distance approach is not efficient (Charcosset et al., 1998). It also has to be noted that the use of BLUP can increase the precision in the evaluation of the general combining ability of inbred parental lines. It may also be used at any step of a breeding programme, in order to better estimate the performance of the individuals.

Use of markers to protect new varieties

We have seen earlier that markers can be used to evaluate the similarity of two individuals at many loci, and the advantage that markers have over phenotypic characterization. This possibility offers perspectives for the protection of newly created varieties (Soller and Beckmann, 1983). This application receives considerable attention, even though the use of molecular markers for this purpose is not yet advocated by the UPOV (Union pour la Protection des Obtentions Végétales). Briefly, if a new variety B is identical to an existing variety A after analysis of many polymorphic loci, there will be a strong presumption that B is in fact a copy of A. It can be noted that to declare B distinct from A on the basis of a difference at a single marker does not constitute sufficient protection to the extent that the genotype at a given locus can be modified relatively easily by backcrossing. It was also proposed during the UPOV convention of 1991 that the concept of "essential derivation" be introduced. This concept is applicable in situations where line B, even though it is distinct from line A, is so close that the economic value of B directly benefits from the selection process of A. An appropriate threshold must be determined for the distance (for example, 25% of the loci at least having fixed different alleles as proposed by Smith et al., 1991). Precision of distance estimation is a key issue for such applications (see above and Table 20).

Perspectives

The applications presented are for the most part used routinely in various selection programmes. Among other potential applications of diversity studies, it has to be noted that the identification of parents from which to develop new breeding populations remains a key issue for most breeders. The value of a population for selection depends not only on its mean, but also on its variance (Gallais, 1990). The average of the population can generally be predicted from the average performance of the parents. Molecular markers have the potential to predict to some extent the variance, which would be particularly useful. First, the properties of the relationship between phenotypic and molecular distances lead to the conclusion that, when developing a breeding population from two lines that are close at the phenotypic level, the variance of the population should be higher for lines that are distant at marker loci than for lines that are similar at marker loci. It also can be shown that, under certain hypotheses, the variance of a quantitative trait within a population is directly related to the diversity of this population for the QTLs involved in its variation. The diversity estimated at the level of marker loci is therefore expected to be predictive of the variance of the population. Some experimental results support the fact that populations developed from parents with a high marker similarity indeed show a low genetic variance (Manjarrez-Sandoval et al., 1997, for soyabean). However, other studies show low correlation coefficients between the genetic variance of populations and the molecular distance between parents (Melchinger et al., 1998, for maize; Kisha et al., 1997, for soyabean; Burkhamer et al., 1998, for spring wheat), so that further investigations are needed to evaluate the potential and limits of this application.

MARKER-ASSISTED SELECTION

Any contribution of molecular markers to selection is based on the availability of markers linked to those genes involved in the variation of selected traits (Stuber, 1995). Thus, the term "marker-assisted selection" refers to any form of selection using markers, for:

- managing the recombinations in order to accumulate favourable genes or chromosome segments in a single genotype and
- reading the genotypic value, at least partly, through the genotype at different marker loci.

These applications can be used for the development or building of particular genotypes as well as for improvement of populations by recurrent selection.

Genotype building

MARKER-ASSISTED BACKCROSSING FOR A MONOGENIC TRAIT

Certain monogenic traits, like susceptibility to diseases and insects, are difficult or expensive to evaluate during cycles of backcrosses. Moreover, in the case of transfer of recessive genes, it is necessary to intercalate one generation of self-fertilization after crossing with the recurrent parent to reveal the recessive homozygotes, which lengthens the transfer process in comparison with that for a dominant gene. Molecular markers offer a solution to these problems. Co-dominant markers allowing the identification of heterozygous genotypes will be easier to use. However, as we have seen in Chapter 3, the use of near-isogenic lines and BSA are methods of choice to obtain markers closely linked to the gene to be introgressed. These markers are generally dominant (RAPD, AFLP, etc.) but they can also be used in marker-assisted backcrosses in cases where the dominant allele of the marker locus is linked to the favourable allele that is to be introduced. Then, the situation becomes similar to that with co-dominant markers, the intercalary self-fertilized generation is no longer necessary. Another advantage of using markers is that individuals can be selected much earlier and without the disturbing effects of the environment. Thus, the use of markers saves time and obviously can offer an economy of means.

Moreover, a major disadvantage of classic backcrossing is the fact that not only the gene of the donor parent, but also a chromosomal fragment surrounding this gene are introduced. Since meiosis hardly breaks up the genome, the length of the donor fragment carrying the selected gene decreases only slowly through the generations. For example, after five backcrosses, it is on average 30 cM for a chromosome of 100 cM (Hanson, 1959; Stam and Zeven, 1981). This was verified experimentally by Young and Tanksley (1989) on different tomato lines in which the gene for resistance to tobacco mosaic Tm-2 was introduced by backcrossing (Fig. 47). Thus, we see that the near-isogenicity of introgressed lines is quite relative. Knowing that there can be several tens of genes per cM on average (and even if all are not polymorphic), it is clear that introgressed lines could differ markedly from

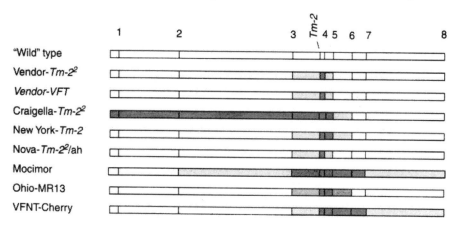

Figure 47. Portion of "donor" genome (in dark gray) introduced by successive backcross around the resistance gene Tm-2 in some varieties of tomato. In light gray: zones of recombination (Young and Tanksley, 1989).

the recurrent line for various traits. Molecular markers also offer a partial solution to this problem, in helping to *restrict the length of the introduced segment*. Besides, they enable us to *accelerate the return to isogenicity* outside this segment. Ideally, there must be two types of markers:

- Those that flank the locus in question and determine a "window" to the right and left in such a way as to detect the closest possible crossover of the gene. It is improbable that the target double recombinant will be found in one generation. Thus, the detection will proceed sequentially, finding a recombination on one side of the gene in BC1 (first backcross generation), and then on the other side in BC2. The probability of these events can be calculated as a function of the population size and the position of markers in the neighbourhood of the gene, which allows us to calculate the population size to be manipulated (Hospital and Charcosset, 1997).
- Those that will allow accelerated return to the recurrent parent and that must thus be distributed over the entire genome. For this, around two markers per chromosome arm will suffice. These markers will mostly be used after BC2 and mainly in BC3. However, they can be used at each previous step, if there are several plants with the favourable recombination event.

In three generations one can thus obtain an isogenicity that is much better than with conventional methods. A theoretical study shows that such a strategy will save around two generations as compared to the traditional approach (Hospital et al., 1992). This is very well illustrated by an experiment on maize by Ragot et al. (1995). The *Bt* gene, which codes for an insecticide protein, was introduced by transgenesis in a line. Genetic analysis showed that *Bt* was located in a single locus on chromosome 1. The objective was thus to transfer it by backcrosses in an elite line. Using markers, the return to the genome of the recurrent parent after three backcrosses proved to be equivalent to that obtained in six generations by conventional selection. Beyond the third backcross, only the region flanking the gene remained heterozygous in the selected individuals (between 2% and 24% of the total length of the chromosome), and after the subsequent backcross it was reduced to maximum of 8.4%, which corresponds to a distance of 14.5 cM.

The same process could be extended to the case of several genes, but the number of generations and/or population size must be increased.

MARKER-ASSISTED BACKCROSSING FOR A POLYGENIC TRAIT

Transfer of a chromosomal segment carrying a QTL
The process described above, or its variants, can be applied to the transfer of a chromosomal segment (which is called QTA, or quantitative trait allele)

identified as favourable during a QTL detection experiment. The donor parent is one of the parents of the cross used for the detection of the QTL, and the recipient parent is the other parent. Apart from the much greater difficulty of phenotypic evaluations, one difficulty arises from the fact that, at each backcross generation, a QTA will be transmitted with a probability less than 1/2, unlike a gene. Indeed, as we have seen in Chapter 4, the confidence interval of the position of a QTL, i.e., the chromosomal segment where the QTL can be, is generally large enough. Thus, during meiosis, there is a probability of recombination within the segment. To keep the QTA, it is necessary to multiply 1/2 by the probability that there will be no recombination within the segment. With a segment of 20 cM marked at the extremities and in the middle, this probability of transmission is around 0.4 instead of 0.5, i.e. $(1 - 0.1)^2$, with the approximation that 10 cM corresponds to $r = 0.1$. Hospital and Charcosset (1997) have shown that using three markers per QTA allows for a good control of the transfer.

An example of QTA transfer using markers in tomato was published by Tanksley et al. (1996). They found QTLs for a series of traits in a BC2 population resulting from a cross between a variety of the cultivated species *Lycopersicon esculentum* (recurrent parent), and a wild species *L. pimpinellifolium* (donor parent). The phenotypic values of each BC2 individual were measured from their BC3 population and from the population resulting from their cross with a tester. Three of the QTLs found, one affecting the fruit weight and two others the fruit shape, were thus independently introgressed to develop near-isogenic lines. Comparison of the lines, after crossing with the tester, showed that in each case the introgression significantly modified the value of the trait in the direction expected from the effect of QTLs measured in BC2. Thus, a QTA of *L. pimpinellifolium* located on chromosome 8 makes the fruits rounder than in the cultivated variety (–38% in an elongation scale). Another QTA for fruit shape, located on chromosome 1, elongates the fruit (+12%). Finally, a QTA of chromosome 9 increases the fruit weight (+8%). It must be noted that, in the last two cases, the QTAs that were manipulated presented an effect opposite to that expected from the parental phenotypes: the fruits of *L. pimpinellifolium* are round and very small, those of *L. esculentum* are large and elongated. It was thus possible to detect in a wild form, which was at first an unlikely bet, QTA improving some "cultivated" traits of the cultivated variety. This method, which the use of markers has made much easier to apply than in the past, undoubtedly is a promising route for marker-assisted selection and exploitation of genetic resources.

Transfer of several chromosomal segments
The principle of backcrossing with several genes can be extended to transfer of several favourable QTAs. In taking into account the probability of non-recombination within the segment, the calculation shows that one

cannot expect to manipulate simultaneously more than four QTAs for population sizes of about 300 (Hospital and Charcosset, 1997). Obviously, this will be easier with greater population sizes and with markers that are very close or even strictly linked to genes.

Schematically, the backcrossing can be done according to the following process:

- Give priority to favourable QTAs of the donor parent. These QTAs, which will always be maintained in the heterozygous state through generations of backcross, must not be lost.
- Select individuals that are carriers of these favourable QTAs for their isogenicity with the recurrent parent outside the manipulated fragments.
- Finish with one or two self-fertilizations to fix the introgressed QTAs.

The experiment of Stuber and Sisco (1993) on grain yield in maize clearly shows the value of marker-assisted genotype building to improve the combining ability of parents of a hybrid. The B73 × Mo17 hybrid was improved by introgression of favourable QTAs of two other lines, Tx303 and Oh43. The search for QTLs has shown the existence of six chromosomal regions for which Tx303 could improve B73, and six others for which Oh43 could improve Mo17. Three backcrosses, followed by two self-fertilizations, were done in each case. The selection on markers (RFLP) was done from the second backcross: the individuals were selected on the basis of the number of segments transferred, and from the percentage of genome of the recurrent parent. The results were encouraging. The crosses between "enhanced B73" lines and Mo17, and between B73 and "enhanced Mo17", gave hybrids that were generally better performing than the control B73 × Mo17, with a gain from 8 to 15%. Analysis of enhanced lines showed that the maximum number of segments transferred was not greater than three, and that the majority contain only a single segment coming from the donor parent: segment 6L for B73 and segment 4S for Mo17. These two QTLs probably have the most significant effect. None of the lines contain the six fragments, which was expected because the probability of obtaining them was very low. Clearly, such an improvement could not have been realized in so short a time using conventional methods.

MARKER-ASSISTED PEDIGREE SELECTION

From an F_2 generation, instead of backcrossing one can attempt to select, among the individuals resulting from self-fertilization (F_3, F_4, etc.), those that have received the best QTAs.

For the purpose of selecting lines on their value per se, one must typically select on the basis of the value of lines that can be derived from one individual. Self-fertilization of a heterozygous individual does not give direct access to this value, which shows the limits of pedigree selection and the advantage of haplodiploidization. The advantage of markers is that they allow us to predict this value, for the marked loci. Let a_l be half the difference between the two homozygotes for the QTL l, as estimated in Chapter 4. For a given trait, the mean value of the lines that can be derived from a genotype i can therefore be written as:

$$\hat{G}^i = \sum_l \hat{g}_l^i$$

(3)

where $\hat{g}_l^i = a_l \theta_l^i$ with $\theta_l^i = -1$ or $+1$ according to whether the genotype at marker considered is homozygous for an allele from an unfavourable parent or from a favourable parent, respectively. The genotypic value of a family from self-fertilization can also be predicted by considering the effects of dominance. An example of the efficiency of this method on tomato is given by Paterson et al. (1991). The efficiency of selection on the basis of markers alone has been compared to that of phenotypic selection, the selection being done from F_2 individuals and the response observed in F_3 families. For the pH of fruits, a trait that is moderately heritable, selection from markers had efficiency comparable to that of phenotypic selection. In contrast, for fruit weight, a trait that is more heritable, phenotypic selection was more effective, while for concentration of soluble sugars, a poorly heritable trait, selection on the basis of markers was the most effective (Table 22). Phenotypic selection is more efficient for a trait with high heritability because it uses the sources of variation of all the QTLs, while markers can use only those due to QTLs to which they are linked. For low heritability, genetic values are better estimated by the markers than by the phenotype, if the population size for QTL detection has been sufficiently large.

In maize, the advantage of selection from markers alone was examined by Stromberg et al. (1994) for the improvement of grain yield. The selection of F_6 families (derived by bulk self-fertilization of F_2 plants) on the basis of only the information from markers linked to QTL detected in F_2 allows us to obtain on average the same progress as with phenotypic selection, while less than 50% of the variation was explained by the markers.

Table 22. Relative efficiency of pedigree selection in F_2 on the basis of values predicted by markers in comparison to selection by phenotype, in a *Lycopersicon esculentum* × *L. chesmanii* cross (Paterson et al., 1991).

	Fruit weight	Fruit pH	Concentration of soluble solids
Heritability	0.45	0.25	0.15
Number of QTLs detected	7	5	4
R^2 (%)	72	34	44
Relative efficiency of markers	0.66	1.07	2.20

If the aim is to select lines for their combining ability with a tester, the effects of QTLs for this aptitude must be estimated. The value of lines that can be derived from an individual will thus be predicted by summation of the effects of each QTL, according to the alleles present, while proceeding as if working with an additive trait (Gallais, 1990).

Marker-assisted recurrent selection

The probability of obtaining the desired genotype in a single cycle of marker-assisted genotype building rapidly decreases when the number of QTAs to be associated increases. Then, it appears better to develop a population in which the frequency of the favourable genotype will be increased through the increase in the frequency of each favourable QTA by several cycles of selection followed by recombination. Such a marker-assisted recurrent selection on markers alone can be developed without or with consideration of QTA effects. However, there is the problem of the unmarked genetic variability, which, for the complex agronomic traits, remains generally quite high (about 50% or more). This variability can only be used within the framework of "conventional" methods of recurrent selection if heritability is sufficient. Thus, to maximize genetic progress it could be efficient to combine both selection on markers and selection on the phenotype in a recurrent scheme. In a first stage such schemes could be applied to F_2 populations, wherever the QTLs have been detected.

RECURRENT SELECTION ON MARKERS ALONE

By a direct selection on the marker giving the same weight to each QTA, the aim of such a method is to increase the frequencies of favourable markers, i.e. markers that are linked to favourable QTAs. Theoretical studies (Luo et al., 1997; Gallais et al., 1997; Hospital et al., 2000) show that the frequencies of selected alleles increase quickly. For example, with 10 independent loci, the frequency of the favourable alleles increases from 0.50 to 0.90 in three generations of selection, and thus the frequency of the favourable genotype increases from 0.001 to 0.35. Then, it could be quite easy to extract the best marker genotype from the improved population. The efficiency of such a method depends mainly on how tightly marker and QTL are linked. When they are not tightly linked, recombination between markers and QTLs quickly decreases the part of the variance explained by the markers. Then, the use of flanking markers will increase efficiency. Considering the rate of fixation and the risk of recombination between markers and their associated QTLs, it will be better to restrict such a method to some selection cycles.

RECURRENT SELECTION ON ADDITIVE VALUE PREDICTED BY MARKERS

To increase the efficiency of recurrent selection schemes, it is important to predict the additive value of individual candidates for selection, the only value that is transmitted from one generation to another. This value is predicted differently according to whether the material is being improved for the purpose of development of hybrid or line varieties.

To prepare the development of hybrid varieties, the material can be improved for its combining ability with a particular tester. As described earlier, the combining ability with the tester can be predicted by a summation of the QTL effects. If the aim is to improve the within-population combining ability, it is necessary to be able to estimate the additive effects. Without the markers, the additive value of an individual is estimated by studying progeny from some mating designs. With markers it is possible to have direct access to this value when the frequency of the marker and the associated effect are known. In the F_2 generation, with a frequency of each marker allele equal to 0.5, the additive effect is related directly to half the difference between the two homozygotes. However, as long as there has been selection, the additive effect of an allele changes, since it depends on its frequency. It can therefore be obtained by the average effect of this allele in combination with all the alleles of the population (Gallais, 1990). This possibility of estimating the additive value at QTL could be a major contribution of molecular markers to recurrent selection. It is a means of calculating, for each individual candidate for selection, the additive value associated with a mapped QTL. The overall additive value is obtained by simple addition of the values predicted at each QTL. If the purpose is to develop line varieties, it is necessary to select for the value of lines that can be derived from a genotype (Gallais, 1990). As in pedigree selection, the advantage of markers is to permit direct access to this value through half the difference between the two homozygotes and summation of the predicted values at each QTL.

The accuracy of the additive values predicted depends on the genetic r^2 values at QTLs detected (% of genetic variance explained by markers), the heritability of the trait, and the population size studied. This size must be larger when the heritability is low. With a sufficiently large size, if the percentage of genetic variation explained by the markers is greater than the narrow-sense heritability, selection on the basis of the predicted values will be more effective, at equal intensity, than selection on phenotype alone (Lande and Thompson, 1990). As selection from markers alone does not require agronomic evaluation, and since the identification of genotypes at markers may be very quick, the duration of a cycle can be reduced to one generation. This may even be done in accelerated conditions, in artificial conditions, or in off-season generations. Then, per unit of time, selection from markers may be more advantageous than phenotypic selection, even when it is less efficient on a cycle basis. Over several cycles of selection the problem arises of recombination between the QTLs detected and their markers. The relative efficiency of the method will thus decline over the course of the cycles. A strong linkage between markers and QTLs is always favourable. The ideal, of course, would be to have available markers in the QTLs themselves. In the absence of such a situation, as with the previous scheme of selection on markers alone, it would be better to restrict its use to only some cycles. It is expected to be more efficient in the short term than the preceding scheme, which gives the same weight to each QTL. The reverse would be true in the long term.

The results of an experiment by Stuber (1989) on an F_2 population of maize showed clearly the possible contribution of markers: a cycle of selection on markers alone was as efficient as phenotypic selection or, for some traits, more effective (Table 23). The experiment also showed that the frequency of favourable marker alleles increased much more with selection on the basis of markers than with phenotypic selection, whereas the progress was the same. Phenotypic selection thus used other sources of variation due to unmarked QTL. It thus seemed that combined selection based on both phenotype and markers would be more effective. The same conclusion could be derived from the experiment by Paterson et al. (1991) on tomato (Table 22).

COMBINED SELECTION BASED ON PHENOTYPE AND MARKERS

Principle and expected efficiency
In several experiments on F_2 populations, selection on the basis of markers alone seemed quite effective, mainly per unit of time. Besides, it does not require agronomic re-evaluation. However, over several generations, its efficiency could decrease very rapidly because of fixation of favourable QTA and recombination between QTLs and markers. It is possible to re-evaluate the relationships between the markers and the improved trait, and then continue with selection on markers alone. Nevertheless, there may be many QTL with effects too weak to be detected, for which variation cannot be used. In simultaneous selection on the basis of phenotypic value and

Table 23. Results of an experiment by Stuber (1989) on grain yield in maize, comparing a cycle of selection on markers alone (after QTL detection) and a cycle of classic phenotypic selection (mass selection).

Type of selection	Response	
	Grain yield	Variation in frequency of favourable QTA
Positive selection on:		
markers alone	151.2	+0.19
phenotype	151.7	+0.06
Negative selection on:		
markers alone	107.7	−0.19
phenotype	122.4	−0.06

markers, all the sources of variation are used, and the final outcome is expected to be better.

In this approach, the markers are to be considered as associated traits: their genetic correlation with the main trait corresponds directly to the square root of the percentage of variance explained by the markers. From general considerations on multitrait selection, it can be predicted that the most effective way of using the two sources of information is to combine them in an index I of the phenotypic value P and the additive value predicted by the markers A (Lande and Thompson, 1990):

$$I = b_1 P + b_2 A \tag{4}$$

where b_1 and b_2 are the coefficients of partial regression of the true genetic value G on P and on A, respectively, which can be derived from the regression theory with the random effects model. It can be demonstrated that:

$$b_1 = \frac{h^2(1 - r_G^2)}{1 - r_P^2} \quad \text{and} \quad b_2 = \frac{1 - h^2}{1 - r_P^2}$$

where h^2 is the narrow-sense heritability, and r_P^2 (or r_G^2) is the fraction of phenotypic (or genotypic) variation explained by the markers. We thus have. $n_P^2 = h^2 r_G^2$ These expressions show that, at fixed r_G^2, the weight given to the phenotypic value increases with the heritability of the trait, while the weight given to the markers decreases.

If the heritability is high, the contribution of molecular markers is small. The advantage of combined selection on the basis of phenotype and markers can thus appear only for fairly low heritability. However, if the heritability becomes too low there will be, for a given number of genotyped plants, more and more errors in the detection of QTL (false positives, poor accuracy on QTL effects). The markers then contribute nothing and could even prove misleading. To keep the advantage of combined selection the population size must be larger when the heritability is low. With only 100 to 300 plants, the heritability that maximizes the relative interest of combined phenotype and marker selection is in the range of 0.15-0.20 (Gallais and Charcosset, 1994; Moreau et al., 1998).

The efficiency of molecular markers is obviously higher when the percentage of genetic variation explained by the markers is high. Actually, it is the product $Nh^2 r_G$ that is to be considered, N being the population size (Lande and Thompson, 1990; Gallais and Rives, 1993). The domain of validity of marker-assisted combined selection corresponds thus to fairly low heritability (between 0.10 and 0.30), fairly high density of markers, and sufficiently large population size (Fig. 48) (Gallais and Charcosset, 1994).

The problem that arises with this type of selection is the cost: it requires a re-evaluation of marker-trait relationships at each cycle. When combined

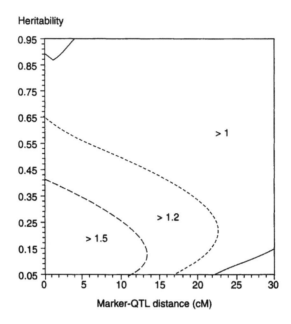

Figure 48. Domain of relative efficiency of marker-assisted selection as a function of heritability and marker-QTL distance, with a population of 300 individuals, 5 QTL with equal effects, and a type I error of 5%. The curves delimit different zones of relative efficiency: > 1.5; 1.2–1.5; 1–1.2 (Moreau et al., 1998).

marker-assisted selection and phenotypic selection are compared at their optimum for the same investment, combined marker-assisted selection seems to have little advantage (Moreau et al., 2000). An efficient solution, justifying the investment in genotyping and experimentation, may consist of alternating combined phenotype + marker selection with selection using markers alone (Gallais et al., 1997; Hospital et al., 1997). Selection on markers alone could be done for two or three cycles in accelerated generations because it does not require agronomic evaluation. Consequently, the progress per unit of time will be higher than with combined selection alone.

Results of simulations
The expected efficiency of combined selection has been verified by simulation by Zhang and Smith (1992), but with large population size, which gives it an advantage even for low heritability. The simulations of Gimelfarb and Lande (1994, 1995) and Hospital et al. (1997) lead to the following observations:

– The progress reaches a plateau much more quickly when the marker-QTL disequilibrium is not revaluated at each generation: the variability explained by markers is quickly vanishing, and better results will be obtained with another subset of markers coming from a new study of marker-trait relationships;

- The effects of the density and number of markers involved in the index seem relatively low compared to that of the population size.
- The method is most advantageous in the case of low heritability.
- The selection must not be too severe, especially when the QTLs are in repulsion.
- An increase in the number of markers per chromosome does not necessarily lead to a more significant response to selection.

COMPARISON WITH CONSTRUCTION OF GENOTYPES

Methods of marker-assisted selection in populations do not directly lead to the development of elite genotypes. However, if selection on the basis of markers alone in populations is compared with marker-assisted' backcrosses, the objective is clearly the same: to assemble in a single genotype the maximum number of favourable alleles. With backcrossing the genotype building is managed by selecting the result of favourable recombinations.   At the level of populations the process is more probabilistic, through increase in the frequency of favourable QTAs. In both cases a population is finally obtained from which the best genotypes must be extracted, which is made simpler by backcrossing.

Similarly, recurrent phenotype + marker selection increases the frequency of associations of favourable genes via the increase in the frequency of marked and unmarked favourable QTAs. Here also, it is necessary to extract the best genotypes, which can be done effectively by haplodiploidization if this technique works in the species considered. But one can of course consider pedigree marker-assisted selection. The advantage of this combined phenotype + marker selection is that it uses all the sources of variation. Crosses between complementary plants can be considered to increase the probabilities of transgression while maintaining the variability at unmarked QTLs. However, with a sufficiently high number of marked QTLs the management of the system is more difficult and costly than intercrossing at random among selected plants.

CHOICE BETWEEN USE OF MARKERS AND INCREASE IN THE NUMBER OF REPLICATIONS

In combined marker-assisted selection, the advantage that markers offer for estimating the genotypic value is equivalent to that of an increase in heritability. Indeed, genetic value is more precisely predicted by markers associated with QTL because each individual carrying a marker of a QTA is a replication of this QTA; the more individuals there are with a given QTA, the greater the precision. Thus, evaluation by markers can be replaced with true replications of individuals, a well-known means to increase heritability. These two means of increasing heritability have been shown to be (Gallais and Charcosset, 1994; Gallais, 1995). The breeder can thus efficient choose between making more replications without using markers, on the one hand, and using the markers and making fewer replications, on the other. The problem is essentially an economic one.

In conclusion about marker-assisted selection, it appears that markers can save time by selection in off-season generations, without agronomic evaluation, and above all they are irreplaceable for the management of recombinations, in order to accumulate favourable alleles as quickly as possible in a single genotype. To exploit their value fully, new schemes of recurrent selection or rather of recurrent genotype construction must be devised.

CONCLUSION

Genetic markers can have many applications in different phases of selection programmes: optimization of genetic resource conservation, choice of parents, identification of favourable alleles, and development of genotypes accumulating such alleles. Some of these applications have already become routine. Analysis of genetic variability for the purpose of organization into groups (or validation of existing groups) and marker-assisted backcrossing are used routinely in a large number of private and government programmes. Other applications remain for the most part in the domain of research. This is particularly the case of selection of quantitative traits for which the main principles have been defined, but for which few experimental results are so far available. There is thus scope for many methodological improvements in optimization of protocols. The future impact of these methods will depend on the results obtained, as well as on the evolution of cost of the informative data (genotyping of an individual for a locus). Nevertheless, markers have already indisputably enriched the range of methods available to the breeder in managing and exploiting genetic variability (Young, 2000; Dekkers and Hospital, 2002).

Appendix 1

PLANT GENETIC MAPS

The following table records the first saturated or partial genetic linkage maps of molecular markers published for the main plant species. For some of them, highly saturated maps now exist, which can be found on websites (http://www.ncbi.nlm.nih.gov/Genomes/index.html, or http://www.agron.missouri.edu/bioservers.html). The markers used were (1) isozymes, (2) proteins, (3) RFLP, (4) RAPD, and (5) STS. The populations used were recombinant lines (RL), hybrid families (F₁), F₂ populations, bulk F₃, first generation backcross (BC), families resulting from open pollination (OP), or doubled haploid (DH). (Table by C. Plomion, INRA Bourdeaux.)

Common name	Species	References	Markers	Populations
Wall cress	*Arabidopsis thaliana*	Chang et al., 1998	3	bulk F_3
		Nam et al., 1989	3	bulk F_3
		Reiter et al., 1992	3, 4	RL
Oat	*Avena* sp.	O'Donoughue et al., 1992	3	bulk F_3
Banana	*Musa acuminata*	Faure et al., 1994	1, 3, 4	F_2
Wheat	*Triticum tauschii*	Chao et al., 1989	3	DH
		Lagudah et al., 1991	3	aneuploid stock
Peanut	*Arachis* sp.	Halward et al., 1993	3	F_2
Sugarcane	*Saccharum* sp.	Da Silva et al., 1993	3	DH
		Al-Janabi et al., 1993	4	DH
Cabbage	*Brassica oleracea*	Slocum et al., 1990	3	F_2
		Landry et al., 1992	3	F_2
		Kianian and Quiros, 1992	3	F_2
Lemon	*Citrus* sp.	Durham et al., 1992	1, 3	BC
		Jarrel et al., 1992	1, 3	F_1
		Cai et al., 1994	3, 4	BC

(Contd...)

(Appendix 1. *Contd...*)

Common name	Species	References	Markers	Populations
Rapeseed	*Brassica napus*	Luro et al., 1995	1, 3, 4	F_1
		Landry et al., 1991	3	bulk F_3
		Ferreira et al., 1994	3	DH
Cucumber	*Cucumis sativus*	Kennard et al., 1994	1, 3, 4	F_2
Eucalyptus	*E. grandis*	Grattapaglia and Sederoff, 1994	4	F_1
	E. urophylla	Grattapaglia and Sederoff, 1994	4	F_1
Broad bean	*Vicia faba*	Torres et al., 1993	1, 3, 4	F_2
Bean	*Phaseolus vulgaris*	Vallejos et al., 1992	1, 2, 3	BC
		Nodari et al., 1993	1, 3, 4	F_2
	Vigna radiata	Menancio-Hautea et al., 1993	3	F_2
	Vigna unguiculata	Menancio-Hautea et al., 1993	3	F_2
Lettuce	*Lactuca sativa*	Landry et al., 1987	1, 3	F_2
		Kesseli et al., 1994	3, 4	F_2
Lentil	*Lens culinaris*	Havey and Muehlbauer, 1989	1, 3	F_2
		Weeden et al., 1992	1, 3	bulk F_3
Alfalfa	*Medicago sativa*	Echt et al., 1994	3, 4	F_1
		Brummer et al., 1993	3	F_2
		Kiss et al., 1993	1, 3, 4	F_2
Maize	*Zea mays*	Helentjaris et al., 1986b	3	monosomics
		Helentjaris et al., 1986a	3	F_2
		Burr et al., 1988	1, 3	RL
		Beavis and Grant, 1991	3	F_2
		Gardiner et al., 1993	1, 3	bulk F_3
		Ahn and Tanksley, 1993	3	RL
		Chao et al., 1994	3	bulk F_3
Millet	*Pennisetum glaucum*	Liu et al., 1994	3	F_2
Mustard	*Brassica rapa*	Song et al., 1991	3	F_2
		Chyi et al., 1992	3	F_2
Blueberry	*Vaccinium* sp.	Rowland and Levi, 1994	4	BC
Barley	*Hordeum vulgare*	Heun et al., 1991	3	DH
		Graner et al., 1991	3	bulk F_3
		Kleinhofs et al., 1993	1, 3, 4	DH
Peach	*Prunus persica*	Chaparro et al., 1994	1, 4	F_2
		Rajapakse et al., 1995	3, 4	F_2

(Contd...)

(Appendix 1. *Contd...*)

Common name	Species	References	Markers	Populations
Petunia	*Petunia hybrida*	Peltier et al., 1994	4	BC
Poplar	*Populus trichocarpa*	Bradshaw et al., 1994	1, 3, 4, 5	F_1
Pine	*Pinus pinaster*	Bahrman and Damerval, 1989	2	OP
		Gerber et al., 1993	2	18 genotypes
		Plomion et al., 1995a	4, 2	OP, F_2
	P. elliottii	Nelson et al., 1993	4	OP
	P. taeda	Devey et al., 1994	1, 3	F_1
Peas	*Pisum sativum*	Ellis et al., 1992	3	RL
		Weeden et al., 1992	1, 3	bulk F_3
		Dirlewanger et al., 1994	3, 4, 5	F_2
Capsicum	*Capsicum annuum*	Tanksley et al., 1988	3	F_2
		Prince et al., 1993	3	F_2
Potato	*Solanum tuberosum*	Bonierbale et al., 1988	3	F_1
		Gebhardt et al., 1989	3	BC
		Gebhardt et al., 1991	3	BC, F_1
		Tanksley et al., 1992	1, 3	BC
Rice	*Oryza sativa*	McCouch et al., 1988	3	F_2
		Saito et al., 1991	3	F_2
		Ahn and Tanksley, 1993	3	BC
		Causse et al., 1994	3	BC
		Inoue et al., 1994	5	F_2
		Kurata et al., 1994	3, 4, 5	F_2
Fir	*Picea glauca*	Tulsieram et al., 1992	4	OP
	P. abies	Binelli and Bucci, 1994	4	OP
Rye	*Secale cereale*	Devos et al., 1993	3	F_2
		Philipp et al., 1994	1, 3, 4	RL
Soybean	*Glycine max*	Keim et al., 1989	3	F_2
		Lark et al., 1993	1, 3	bulk F_3
Sorghum	*Sorghum bicolor*	Hulbert et al., 1990	3	F_2
		Binelli et al., 1992	3	F_2
		Whitkus et al., 1992	3	F_2
		Pammi et al., 1994	4	F_2
		Ragab et al., 1994	3	bulk F_3
Tomato	*Lycopersicon esculentum*	Helentjaris et al., 1986a	3	F_2
		Bernatzky and Tanksley, 1986	1, 3	F_2, BC
		Bonierbale et al., 1988	3	F_2
		Tanksley et al., 1988	1, 3	F_2, BC
		Tanksley et al., 1992	1, 3	F_2
		Eshed et al., 1992	3	BC
		Paran et al., 1995	3	RL

Appendix 2

COMPILATION OF FIRST QTLs MAPPED IN PLANTS

The number in parentheses next to the trait indicates the number of components of the trait analysed. The *Material and methods* column indicates the number of individuals analysed, the type of population used, the number of markers (mk), and the detection method used (ML, interval mapping by the maximum likelihood method; AN, ANOVA). The *Results* column presents the minimum and maximum values of numbers of QTLs detected and the percentages of total variation (R^2tot) and individual variation (R^2ind) for all the components of the trait. These data were compiled prior to 1996. For recent compilations, see for example http://arsgenome.cornell.edu or http://www.agron.missouri.edu/bioservers.html.

Plant	Trait	Material and methods	Results	References
Maize	Yield (10)	1700 F_2; 20 mk; AN	R^2tot: 19-39%; R^2ind: 0.2-1.3%	Edwards et al., 1987
	Yield (4)	1930 F_2; 20 mk; AN	R^2ind: 0.3-3%	Stuber et al., 1987
	Height	112 $F_{2:4}$; 148 mk; ML 112 $F_{2:4}$; 106 mk; ML 144 $F_{2:4}$; 78 mk; ML 144 $F_{2:4}$; 68 mk; ML	6 QTL; R^2tot: 73% 4 QTL; R^2tot: 53% 3 QTL; R^2tot: 34% 3 QTL; R^2tot: 45%	Beavis et al., 1991
	Domestication (11)	260 F_2; 58 mk; AN	4-7 QTL; R^2tot: 34-95%; R^2ind: 5-78%	Doebley and Stec, 1991
	Low phosphorus stress	90 $F_{2:3}$; 77 mk; AN	6 QTL; R^2tot: 46%	Reiter et al., 1991
	Pollen (2)	45 RIL; 200 mk; AN	5-6 QTL; R^2tot: 79-89%; R^2ind: 10-27%	Sari-Gorla et al., 1992
	Yield (12)	220 F_2 × P; 34 mk; AN	2-11 QTL; R^2tot: 9-48%; R^2ind: 2-11%	Zehr et al., 1992
	Yield	264 $F_{2:3}$; × P1 & P2; 76 mk; ML	6-8 QTL; R^2tot: 59-61%; R^2ind: 6-18%	Stuber et al., 1992
	Domestication (10)	290 F_2; 82 mk; ML	2-7 QTL; R^2tot: 42-87%; R^2ind: 4-42%	Doebley and Stec, 1993

Protein & amino acids (2)	100 $F_{2:3}$; 100 mk; AN	9-10 QTL; R^2tot: 78-84%	Goldman et al., 1993
Resistance to ear rot	150 $F_{2:3}$; 105 mk; ML	5 QTL; R^2tot: 20%; R^2ind: 4-9%	Pe et al., 1993
Resistance to gray leaf spot	337 $F_{2:3}$; 87 mk; AN	15 QTL; R^2tot: 2-15%	Bubeck et al., 1993
Resistance to Exerohilum turcicum (3)	150 $F_{2:3}$; 103 mk: ML	2-5 QTL; R^2tot: 29-45%; R^2ind: 7-18%	Freymark et al., 1993
Resistance to corn borer (2)	300 $F_{2:3}$; 80 mk; ML	3-7 QTL; R^2tot: 38-63%; R^2ind: 3-34%	Schon et al., 1993
Heat shock proteins (5)	45 RIL; 200 mk; AN	3-8 QTL; R^2tot: 35-60%	Frova and Sari-Gorla, 1993
Yield (17)	112 $F_{2:3}$ & F_4; 90 mk; ML	0-6 QTL; R^2ind: 6-27%	Beavis et al., 1994
Yield (3)	380 $F_{2:3}$ × T1 & T2; 89 mk: ML	4-8 QTL; R^2tot: 32-60%; R^2ind: 4-33%	Schon et al., 1994
Morphology (5)	150 $F_{2:3}$; 130 mk; ML	2-6 QTL; R^2tot: 61-81%; R^2ind: 6-53%	Veldboom and Lee, 1994
Lipids (2)	200 $F_{2:3}$; 80 mk; AN	10-13 QTL; R^2tot: 43-47%	Goldman et al., 1994
Quantity of individual proteins (72)	60 F_2; 109 mk; ML	0-5 QTL; R^2tot: 37-90%; R^2ind: 14-67%	Damerval et al., 1994
Yield (8)	387 F_2; 46 mk; ML	2-5 QTL; R^2ind: 4-36%	Ragot et al., 1995
Yield (3)	232 $F_{2:3}$ × T1 & T2; 72 mk; ML	2-6 QTL; R^2tot: 22-41%; R^2ind: 6-20%	Ajmone-Marsan et al., 1995
Composition of grain (7)	200 $F_{2:3}$; 80 mk; AN	8-13 QTL; R^2tot: 22-52%	Alferai et al., 1995
Carbon metabolism (10)	65 RIL; 109 mk; AN	3-6 QTL; R^2ind: 7-32%	Causse et al., 1995
Yield (8)	186 RIL; 101 mk; AN	5-12 QTL; R^2tot: 17-53%; R^2ind: 2-10%	Austin and Lee, 1996
Resistance to gray leaf spot (6)	239 $F_{2:3}$; 78 mk; ML	3 QTL; R^2ind: 7-56%	Saghai-Maroof et al., 1996
Drought tolerance (4)	234 $F_{2:3}$; 142 mk; ML	4-8 QTL; R^2tot: 33-50%; R^2ind: 5-15%	Ribaut et al., 1996
Wheat	Grain hardiness	114 RIL; 1100 mk; AN	5 QTL; R^2ind: 4-63%	Sourdille et al., 1996
Millet	Resistance to oidium (4)	119 $F_{2:4}$; 22 mk; ML	2-4 QTL; R^2tot: 32-65%; R^2ind: 8-47%	Jones et al., 1995

(Contd...)

(Appendix 2. *Contd...*)

Plant	Trait	Material and methods	Results	References
Sorghum	Height, flowering	370 F_2; 71 mk; ML	3-6 QTL; R^2ind: 4-86%	Lin et al., 1995
	Height	152 F_2; 111 mk; ML	4 QTL; R^2tot: 63%; R^2ind: 9-29%	Pereira and Lee, 1995
Rice	Grain elongation	85 F_3; 170 mk; ML	1 QTL; R^2: 14%	Ahn et al., 1993b
	Yield (12)	195 RIL × (P1 & P2); 141 mk; ML	1-5 QTL; R^2tot: 0-42%	Xiao et al., 1996b
	Height, flowering	241 F_4; 113 mk; ML	3-4 QTL; R^2tot: 53-79%; R^2ind: 8-45%	Li et al., 1995
	Resistance to *Rhizoctonia solani*	255 F_4; 113 mk; AN	6 QTL; R^2tot: 47%	Li et al., 1995
	Drought tolerance (4)	203 RIL; 127 mk; AN	8-18 QTL; R^2tot: 19-57%; R^2ind: 6-34%	Champoux et al., 1995
	Drought tolerance (4)	202 RIL; 125 mk; ML	4-19 QTL; R^2tot: 13-34%; R^2ind: 6-19%	Ray et al., 1996
	Yield (13)	195 RIL × (P1 & P2); 141 mk; ML	1-6 QTL; R^2tot: 7-74%	Xiao et al., 1996b
	Yield (8)	171 F_2; 93 mk; ML 170 F_2; 101 mk; ML	0-3 QTL; R^2ind: 8-23%	Lin et al., 1996
	Germination	204 $F_{2:3}$; 117 mk; ML	2-5 QTL; R^2ind: 7-38%	Redona and McKill, 1996
Barley	Resistance to oidium	113 DH; 159 mk; ML	2 QTL; R^2tot: 20%; R^2ind: 10-12%	Heun, 1992
	Yield (8)	150 DH; 123 mk; ML	6-10 QTL; R^2tot: 56-72%	Hayes et al., 1993
	Cold tolerance	100 DH; 78 mk; ML	1-5 QTL; R^2tot: 48-79%; R^2ind: 11-79%	Han et al., 1996
	Resistance to rust (2)	100 DH; 78 mk; AN	2 QTL; R^2ind: 10-57%	Chen et al., 1994
	Yield (13)	59 DH; 83 mk; ML	2-10 QTL	Thomas et al., 1995
	Resistance and develop-ment (8)	250 DH; 50 mk; ML	1-5 QTL; R^2tot: 5-52%	Backes et al., 1995
	Resistance to *P. teres* and *C. sativus*	150 DH; 123 mk; ML	1-7 QTL; R^2tot: 67-70%	Steffenson et al., 1996
	Height, precocity	99 DH; 99 mk; AN	3-8 QTL	Bezant et al., 1996

Tomato	Fruit quality (3)	237 BC_1; 70 mk; ML	4-6 QTL; R^2tot: 44-58%	Paterson et al., 1988
	Fruit quality (3)	350 $F_{2.3}$; 71 mk; ML	7-13 QTL; R^2tot: 44-72%; R^2ind: 4-42%	Paterson et al., 1991
	Morphology (10)	432 F_2; 98 mk; ML	3 10 QTL; R^2tot: 18-61%; R^2ind: 3-34%	de Vicente and Tanksley, 1993
	Resistance to TYLC virus	50 BC_1S_1; 61 mk; AN	3 QTL	Zamir et al., 1994
	Yield (5)	49 NIL; 375 mk; AN	11-23 QTL	Eshed and Zamir, 1995
	Fruit quality (3)	97 RIL; 132 mk; AN	7-13 QTL; R^2ind: 4-42%	Goldman et al., 1995
	Morphology (19)	257 BC_1; 120 mk; ML	2-7 QTL; R^2tot: 12-58%; R^2ind: 4-44%	Grandillo and Tanksley 1996
	Morphology (21)	170 BC_2; 121 mk; AN	2-12 QTL; R^2ind: 4-37%	Tanksley et al., 1996
	Acyl-sugars (4)	196 F_2; 150 mk; AN	5 QTL; R^2ind: 8-20%	Mutschler et al., 1996
Potato	Resistance to insects (4)	150 BC_1; 80 mk; ML 150 BC_2; 70 mk; ML	2 QTL; R^2tot: 52-63%	Bonierbale et al., 1993
	Resistance to nematodes	96 F'_1; 29 mk; AN	2 QTL; R^2tot: 14%	Kreike et al., 1993
	Dormancy	110 F'_1; 117 mk; AN	6 QTL; R^2tot: 57%; R^2ind: 4-20%	Freyre et al., 1994
	Resistance to P. infestans (2)	189 F'_1; 111 mk; AN	11 QTL	Leonards-Schippers et al., 1994
	Tuberization and dormancy	150 BC_1; 80 mk; ML 150 BC_2; 70 mk; ML	2-5 QTL; R^2ind: 5-31%	van den Berg et al., 1996
	Yield and resistance to nematodes	80 BC_1; 68 mk; ML	0-2 QTL; R^2ind: 16-41%	Kreike et al., 1996
Chilli	Resistance to P. infestans (4)	94 DH; 119 mk; AN	0-4 QTL; R^2tot: 54-73%; R^2ind: 17-55%	Lefebvre and Palloix, 1996
Soybean	Morphology	60 $F_{2.3}$; 150 mk; AN	1-5 QTL; R^2ind: 12-24%	Keim et al., 1990
	Yield (10)	69 F_5; 132 mk; ML	1-3 QTL; R^2ind: 17-53%	Mansur et al., 1993a
	Resistance to nematodes	56 F_2; 43 mk; ML	2 QTL; R^2tot: 52%; R^2ind: 21-40%	Concibido et al., 1996
	Grain weight (2)	150 $F_{2.3}$; 91 mk; AN	3-5 QTL; R^2tot: 50-60%	Maughan et al., 1996a

(Contd...)

(Appendix 2. *Contd...*)

Plant	Trait	Material and methods	Results	References
	Development (3)	111 F_2; 154 mk; ML	2-3 QTL; R^2ind: 8-64%	Lee et al., 1996
	Resistance to nematodes	200 $F_{2:3}$; 56 mk; AN	4 QTL; R^2ind: 7-91%	Vierling et al., 1996
Vigna	Grain weight	58 F_2; 104 mk; ML	4 QTL; R^2tot: 50%	Fatokun et al., 1992
	Resistance to mildew	58 F_2; 145 mk; AN	3 QTL; R^2tot: 58%	Young et al., 1993
Peas	Resistance and development	72 F_2; 69 mk; AN	1-5 QTL; R^2tot: 19-75%; R^2ind: 10-46%	Dirlewanger et al., 1994
	Grain weight	102 F_2; 199 mk; ML 51 RIL; 235 mk; ML	3-7 QTL; R^2ind: 13-26%	Timmerman-Vaughan et al., 1996
Bean	Resistance to CCB (2)	70 $F_{2:3}$; 152 mk; ML	4 QTL; R^2tot: 52-75%; R^2ind: 11-32%	Nodari et al., 1993
Brassica	Morphology (6)	90 F_2; 72 mk; AN	2-5 QTL; R^2tot: 31-60%; R^2ind: 8-39%	Kennard et al., 1994
	Flowering (3)	92 $F_{2:3}$; 112 mk; ML	2-3 QTL; R^2tot: 54%; R^2ind: 16-29%	Camargo and Osborn, 1996
	Palmitic acid	105 F_2; 140 mk; AN	1 QTL; R^2: 55%	Tanhuanpaa et al., 1995
	Oleic acid	100 F_2; 6 mk; ML	1 QTL; R^2: 53%	Tanhuanpaa et al., 1996
	Lipids (2)	151 DH; 205 mk; ML	2-3 QTL; R^2tot: 51%	Ecke et al., 1995
	Linoleic acid and erucic acid	99 DH; 199 mk; ML	2 QTL; R^2tot: 60-95%	Thormann et al., 1996
Cucumber	Fruit quality (6)	100 $F_{2:3}$; 69 mk; ML	1-5 QTL; R^2tot: 12-64%	Kennard et al., 1995
Poplar	Growth	55 $F_{2:3}$; 343 mk; ML	1-5 QTL; R^2tot: 32-85%	Bradshaw and Stettler, 1995
Cacao	Morphology (4)	131 BC_1; 138 mk; ML	2-4 QTL; R^2tot: 11-25%; R^2ind: 5-14%	Crouzillat et al., 1996
Eucalyptus	Vegetative multiplication (3)	62 F_1; 100 mk; ML	4-10 QTL; R^2tot: 8-41%; R^2ind: 6-21%	Grattapaglia et al., 1995

References

Aagaard, J.E., Krutovskii, K.V., and Strauss, S.H. 1998. RAPDs and allozymes exhibit similar levels of diversity and differentiation among populations and races of Douglas-fir. *Heredity*, 81, 69-78.

Adams, M.D., Kelley, J.M., Gocayne, J.D., Dubnick, M., Polymeropoulos, M.H., Xiao, H., Merril, C.R., Wu, A., Olde, B., Moreno, F., Kerlavage, A.R., McCombie, R.W., and Venter, J.C. 1991. Complementary DNA sequencing: expressed sequence tags and human genome project. *Science*, 252, 1651-1656.

Adams, W.T. 1992. Gene dispersal within forest tree populations. *New Forests*, 6, 217-240.

Ahn, S., Anderson, J.A., Sorrells, M.E., and Tanksley, S.D. 1993a. Homoeologous relationships of rice, wheat and maize chromosomes. *Mol. Gen. Genet.*, 241, 483-490.

Ahn, S., and Tanksley, S.D. 1993. Comparative linkage maps of the rice and maize genomes. *Proc. Natl. Acad. Sci. USA*, 90, 7980-7984.

Ahn, S.N., Bollich, C.N., McClung, A.M., and Tanksley, S.D. 1993b. RFLP analysis of genomic regions associated with cooked-kernel elongation in rice. *Theor. Appl. Genet.*, 87, 27-32.

Ajmone-Marsan, P., Castiglioni, P., Fusari, F., Kuiper, M. and Motto, M. 1998. Genetic diversity and its relationship to hybrid performance in maize as revealed by RFLP and AFLP markers. *Theor. Appl. Genet.*, 96, 219-227.

Ajmone-Marsan, P., Monfredini, G., Ludwig, W.F., Melchinger, A.E., Franceschini, P., Pagnotto, G., and Motto, M. 1995. In an elite cross of maize a major quantitative trait locus controls one-fourth of the genetic variation for a grain yield. *Theor. Appl. Genet.*, 90, 415-424.

Alferai, R., Berke, T.K., and Rocheford, T.R. 1995. Quantitative trait locus analysis of fatty acid concentrations in maize. *Genome*, 38, 894-901.

Al-Janabi, S.M., Honeycutt, R.J., McClelland, M., and Sobral, B.W.S. 1993. A genetic linkage map of *Saccharum spontaneum* L. "SES 208". *Genetics*, 134, 1249-1260.

Allard, R.W. 1956. Formulas and tables to facilitate the calculation of recombination values in heredity. *Hilgardia*, 24, 235-278.

Arnheim, N., Strange, C., and Erlich, H. 1985. Use of pooled DNA samples to detect linkage disequilibrium of polymorphic restriction fragments and human disease, Studies of the HLA class II loci. *Proc. Natl. Acad. Sci. USA*, 92, 6970-6974.

Arondel, V., Lemieux, B., Hwang, I., Gibson, S., Goodmann, H.M., and Somerville, C.R. 1992. Map-based cloning of a gene controlling omega-3 fatty acid desaturation in *Arabidopsis*. *Science*, 258, 1353-1355.

Arumuganathan, K., and Earle, E.D. 1991. Nuclear DNA content of some important plant species. *Plant. Mol. Biol. Rep.*, 9, 208-218.

Ashley, M.V., and Dow, B.D. 1994. The use of microsatellite analysis in population biology: background, methods and potential applications. In *Molecular Ecology and Evolution*, B. Schierwater, B. Streit, G.P. Wagner, R. DeSalle, eds. Birkhauser, Verlag, pp. 185-201.

Asins, M.J., and Carbonell, E.A. 1988. Detection of linkage between restriction fragment length polymorphism markers and quantitative traits. *Theor. Appl. Genet.*, 76, 623-626.

Asnaghi, C., Paulet, F., Kaye, C., Grivet, L., Deu, M., Glaszmann, J.C., and D'Hont A. 2000. Application of synteny across Poaceae to determine the map location of a sugarcane rust resistance gene. *Theor. Appl. Genet.*, 101, 962-969.

Austerlitz, F., and Smouse, P.E. 2001. Two-generation analysis of pollen flow across a landscape. II. Relation between Φ_{ft}, pollen dispersal and interfemale distance. *Genetics* 157, 851-857.

Austin, D.F., and Lee, M. 1996. Comparative mapping in $F_{2:3}$ and $F_{6:7}$ generations of quantitative trait loci for grain yield and yield components in maize. *Theor. Appl. Genet.*, 92, 817-826.

Ayres, D.R., and Ryan, F.J. 1997. The clonal and population structure of a rare endemic plant, *Wyethia reticulata* (Asteraceae): allozyme and RAPD analysis. *Mol. Ecol.*, 6, 761-772.

Ayres, D.R., and Ryan, F.J. 1999. Genetic diversity and structure of the narrow endemic *Wyethia reticulata* and its congener *W. bolanderi* (Asteraceae) using RAPD and allozyme techniques. *Amer. J. Bot.*, 86, 344-353.

Backes, G., Graner, A., Foroughi-Wehr, B., Fischbeck, G., Wenzel, G., and Jahoor, A. 1995. Localization of quantitative trait loci (QTL) for agronomically important characters by the use of a RFLP map in barley (*Hordeum vulgare* L.). *Theor. Appl. Genet.*, 90, 294-302.

Bahrman, N., and Damerval, C. 1989. Linkage relationships of loci controlling protein amounts in maritime pine (*Pinus pinaster* Ait.). *Heredity*, 63, 267-274.

Bailey, N.J.T. 1961. *Introduction to the Mathematical Theory of Genetic Linkage*. Oxford, Clarendon Press, 294 pp.

Bar-hen, A., and Charcosset, A. 1994. Relationship between molecular and morphological distances in a maize inbred lines collection. Application for breeders' rights protection. In *Biometrics in Plant Breeding: Applications of Molecular Markers*. Proc. 9th Meeting of the Eucarpia

Section Biometrics in Plant Breeding (eds. J.W. van Oijen and J. Jansen), pp. 57-66.

Bark, O.H., and Havey, M.J. 1995. Similarities and relationships among populations of the bulb onion as estimated by RFLPs. *Theor. Appl. Genet.*, 90, 407-414.

Barker, J.S.F., Moore, S.S., Hetzel, D.J.S., Evans, D., Tan, S.G., and Byrne, K. 1997a. Genetic diversity of Asian water buffalo (*Bubalus bubalis*): microsatellite variation and a comparison with protein-coding loci. *Animal Genet.*, 28, 103-115.

Barker, J.S.F., Tan, S.G., Selvaraj, O.S., and Mukherjee, T.K. 1997b. Genetic variation within and relationships among populations of Asian water buffalo (*Bubalus bubalis*). *Animal Genet.*, 28, 1-13.

Baron, H., Fung, S., Aydin, A., Bahring, S., Luft, F.C., and Schuster, H. 1996. Oligonucleotide ligation assay (OLA) for the diagnosis of familial hypercholesterolemia. *Nat. Biotechnol.*, 14, 1279-1282.

Baruffi, L., Damiani, G., Guglielmino, C.R., Bandi, C., Malacrida, A.R., and Gasperi G. 1995. Polymorphism within and between populations of *Ceratitis capitata*: comparison between RAPD and multilocus enzyme electrophoresis data. *Heredity*, 74, 425-437.

Beavis, W.D., and Grant, D. 1991. A linkage map based on information from F_2 populations of maize (*Zea mays* L.). *Theor. Appl. Genet.*, 82, 636-644.

Beavis, W.D., Grant, D., Albertsen, M., and Fincher, R. 1991. Quantitative trait loci for plant height in four maize populations and their associations with qualitative genetic loci. *Theor. Appl. Genet.*, 83, 141-145.

Beavis, W.D., and Keim, P. 1996. Identification of quantitative trait loci that are affected by environment. In *Genotype by Environment Interaction*, Kang, M.S., Gauch, H.G., eds. CRC Press Inc., pp. 123-149.

Beavis, W.D., Smith, O.S., Grant, D., and Fincher, R. 1994. Identification of quantitative trait loci using a small sample of topcrossed and F_4 progeny from maize. *Crop Sci.*, 34, 882-896.

Becker, J., Vos, P., Kuiper, M., Salamini, F., and Heun, M. 1995. combined mapping of AFLP and RFLP markers in barley. *Mol. Gen. Genet.*, 249, 65-73.

Beismann, H., Barker, H.A., Karp, A., and Speck, T. 1997. AFLP analysis sheds light on distribution of two *Salix* species and their hybrid along a natural gradient. *Mol. Ecol.*, 6, 989-993.

Bell, C.J., and Ecker, J.R. 1994. Assignment of thirty microsatellite loci to the linkage map of *Arabidopsis*. *Genomics*, 19, 137-144.

Bennett, M.D., and Smith, J.B. 1991. Nuclear DNA amount in angiosperms. *Proc. R. Soc. Lond. B*, 334, 309-345.

Bennetzen, J.L., and Freeling, M. 1993. Grasses as a single genetic system: genome composition, colinearity and compatibility. *Trends Genet.*, 9, 259-261.

Bernacchi, D., Beck-Bunn, T., Emmatty, D., Eshed, Y., Inai, S., Lopez, J., Petiard, V., Sayama, H., Uhlig, J., Zamir, D., and Tanksley, S. 1998a. Advanced backcross QTL analysis of tomato. II. Evaluation of near-isogenic lines carrying single-donor introgressions for desirable wild QTL-alleles derived from *Lycopersicon hirsutum* and *L. pimpinellifolium*. *Theor. Appl. Genet.*, 97, 170-180.

Bernacchi, D., Beck-Bunn, T., Eshed, Y., Lopez, J., Petiard, V., Uhlig, J., Zamir, D., and Tanksley, S. 1998b. Advanced backcross QTL analysis in tomato. I. Identification of QTLs for traits of agronomic importance from *Lycopersicon hirsutum*. *Theor. Appl. Genet.*, 97, 381-397.

Bernardo, R. 1993. Estimation of coefficient of coancestry using molecular markers in maize. *Theor. Appl. Genet.*, 85, 1055-1062.

Bernardo, R. 1994. Prediction of maize single cross performance using RFLPs and information from related hybrids. *Crop Sci.*, 34, 20-25.

Bernardo, R., and Kahler, A.L. 2001. North American study on essential derivation in maize: inbreds developed without and with selection from F_2 populations. *Theor. Appl. Genet.*, 102, 986-992.

Bernatzky, R., and Tanksley, S.D. 1986. Towards a saturated linkage map in tomato based on isozymes and random cDNA sequences. *Genetics*, 112, 887-898.

Bezant, J., Laurie, D., Pratchett, N., Chojecki, J., and Kearsey, M. 1996. Marker regression mapping of QTL controlling flowering time and plant height in a spring barley (*Hordeum vulgare* L.) cross. *Heredity*, 77, 64-73.

Bhattacharyya, M.K., Smith, A.M., Ellis, T.H.N., Hedley, C., and Martin, C. 1990. The wrinkled seed character of pea described by Mendel is caused by a transposon-like insertion in a gene encoding starch-branching enzyme. *Cell*, 60, 115-122.

Binelli, G., and Bucci, G. 1994. A genetic linkage map of *Picea abies* Karts, based on RAPD markers, as a tool in population genetics. *Theor. Appl. Genet.*, 88, 283-288.

Binelli G., Gianfranceschi, L., Pe, M.E., Taramino, G., Busso, C., Stenhouse, J., and Ottaviano, E. 1992. Similarity of maize and sorghum genomes as revealed by maize RFLP probes. *Theor. Appl. Genet.*, 84, 10-16.

Birky, C., Jr. 1988. Evolution and variation in plant chloroplast and mitochondrial genomes. In *Plant Evolutionary Biology*, L.D. Gottlieb and S.K. Jain, eds., Sinauer Associates, pp. 21-46.

Birky, C., Jr. 1991. Evolution and population genetics of organelle genes: mechanisms and models. In *Evolution at the Molecular Level*, R.K. Selander, A.G. Clark and T.S. Whittam, eds., Sinauer Associates, pp. 112-134.

Birky, C.W., Fuerst, P., and Maruyama, T. 1989. Organelle gene diversity under migration, mutation and drift: equilibrium expectations, approach to equilibrium, effects of heteroplasmic cells, and comparison to nuclear genes. *Genetics*, 121, 613-627.

Blanc G., Barakat A., Guyot R., Cooke R., Delseny M. 2000. Extensive duplication and reshuffling in the Arabidopsis genome. *Plant Cell*, 12, 1093-1101.

Bodenes, C., Labbe, T., Pradere, S., and Kremer, A. 1997. General vs. local differentiation between two closely related oak species. *Mol. Ecol.*, 6, 713-724.

Bonierbale, M.W., Plaisted, R.L., Pineda, O., Tanksley, S.D. 1993. QTL analysis of trichome mediated insect resistance in potato. *Theor. Appl. Genet.*, 87, 973-987.

Bonierbale, M.W., Plaisted, R.L., and Tanksley, S.D. 1988. RFLP maps based on a common set of clones reveal modes of chromosomal evolution in potato and tomato. *Genetics*, 120, 1095-1103.

Boppenmeier, J., Melchinger, A.E., Brunklaus-Jung, E., Geiger, H.H., and Herrmann, R.G. 1992. Genetic diversity for RFLPs in European maize inbreds: I. relation to performance of Flint × Dent crosses for forage traits. *Crop. Sci.*, 32, 895-902.

Borts, R.H., and Haber, J.E. 1987. Meiotic recombination in yeast: alteration by multiple heterozygosities. *Science*, 237, 1459-1465.

Bost B., de Vienne D., Hospital F., Moreau L., Dillmann C. 2001. Genetic and nongenetic bases for the L-shaped distribution of quantitative trait loci effects. *Genetics*, 157, 1773-1787.

Bost B., Dillmann C., de Vienne D. 1999. Fluxex and metabolic pools as model traits for quantitative genetics. I. The L-shaped distribution of gene effects. *Genetics*, 153, 2001-2012.

Botstein, D., White, R.L., Skolnick, M., and Davis, R.W. 1980. Construction of a genetic linkage map in man using restriction fragment length polymorphisms. *Am. J. Hum. Genet.*, 32, 314-331.

Bousquet, J., Strauss, S.H., Doerksen, A.H., and Price, R.A. 1992. Extensive variation in evolutionary rate of *rbcL* gene sequences among seed plants. *Proc. Natl. Acad. Sci. USA*, 89, 7844-7848.

Bradshaw, H.D., and Stettler, R.F. 1995. Molecular genetics of growth and development in *Populus*. IV. Mapping QTLs with large effects on growth, form and phenology traits in a forest tree. *Genetics*. 139, 963-973.

Bradshaw, H.D., Villar, M., Watson, B.D., Otto, K.G., Stewart, S., and Stettler, R.F. 1994. Molecular genetics of growth and development in *Populus*. III. A genetic linkage map of a hyrid poplar composed of RFLP, STS, and RAPD markers. *Theor. Appl. Genet.*, 89, 167-178.

Breto, M.P., Asins, M.J., and Carbonell, E.A. 1994. Salt tolerance in *Lycopersicon* species. III. Detection of QTLs by means of molecular markers. *Theor. Appl. Genet.*, 88, 395-401.

Brown, A.H.D. 1989. Core collections: a practical approach to genetic resources management. *Genome*, 31, 318-324.

Brown, A.H.D., Burdon, J.J., and Jarosz, A.M. 1989. Isozyme analysis of plant mating system. In *Isozyme in Plant Biology*. Soltis, D.E. and Soltis, P.S., eds., pp. 73-105.

Brown, A.H.D., and Munday, J. 1982. Population genetic structure and optimal sampling of land races of barley from Iran. *Genetica*, 58, 85-96.

Brown, A.H.D., and Weir, B.S. 1983. Measuring genetic variability in plant populations. In *Isozymes in Plant Genetics and Breeding*, Part A, S.D. Tanksley and T.J. Orton, eds., Elsevier, Amsterdam.

Bruford, M.W., and Wayne, R.K. 1993. Microsatellites and their application to population genetic studies. *Cur. Op. Genet. Dev.*, 3, 939-943.

Brummer, E.C., Bouton, J.H., and Kochert, G. 1993. Development of an RFLP map in diploid alfalfa. *Theor. Appl. Genet.*, 86, 329-332.

Bubeck, D.M., Goodman, M.M., Bevis, W.D., and Grant, D. 1993. Quantitative trait loci controlling resistance to gray leaf spot in maize. *Crop Sci.*, 33, 838-847.

Bucci, G., Vendramin, G.G., Lelli, L., and Vicario, F. 1997. Assessing the genetic divergence of *Pinus leucodermis* Ant. endangered populations: use of molecular markers for conservation purposes. *Theor. Appl. Genet.*, 95, 1138-1146.

Burke, D.T., Carle, G.F., and Olson, M.V. 1987. Cloning of large segments of exogenous DNA into yeast by means of artificial chromosome vectors. *Science*, 236, 806-812.

Burkhamer, R.L., Lanning, S.P., Martens, R.J., Martin, J.M., and Talbert, L.E. 1998. Predicting progeny variance from parental divergence in hard red spring wheat. *Crop Sci.*, 38, 243-248.

Burr, B, and Burr, F.A. 1991. Recombinant inbreds for molecular mapping in maize. *Trends Genet.*, 7, 55-60.

Burr, B., Burr, F.A., Matx, E.C., and Romero-Severson, J. 1992. Pinning down loose ends: mapping telomeres and factors affecting their length. *The Plant Cell*, 4, 953-960.

Burr, B., Burr, F.A., Thompson, K.H., Albertsen, M.C., and Stuber, C.W. 1988. Gene mapping with recombinant inbreds in maize. *Genetics*, 118, 519-526.

Burstin, J., and Charcosset, A. 1997. Relationship between phenotypic and marker distances: theoretical and experimental investigations. *Heredity*, 79, 477-483.

Burstin, J., Charcosset, A., Barriere, Y., Hebert, Y., de Vienne, D., and Damerval, C. 1995. Molecular markers and protein quantities as genetic descriptors in maize. II. Prediction of performance of hybrids for forage traits. *Plant Breeding*, 114, 427-433.

Buso, G.S.C., Rangel, P.H., and Ferreira, M.E. 1998. Analysis of genetic variability of South American wild rice populations (*Oryza glumaepatula*) with isozymes and RAPD markers. *Mol. Ecol.*, 7, 107-117.

Byrne, M., and Moran, G.F. 1994. Population divergence in the chloroplast genome of *Eucalyptus nitens*. *Heredity*, 73, 18-28.

Byrne, P.F., McMullen, M.D., Snook, M.E., Musket, T.A., Theuri, J.M., Widstrom, N.W., Wiseman, B.R., and Coe, E.H. 1996. Quantitative trait loci and metabolic pathways: genetic control of the concentration of maysin, a corn earworm resistance factor, in maize silks. *Proc. Natl. Acad. Sci. USA*, 93, 8820-8825.

Caetano-Anolles, G. 1994. MAAP: a versatile and universal tool for genome analysis. *Plant Mol. Biol.*, 25: 1011-1026.

Caetano-Anolles, G., Bassam, B.J., and Gresshoff, P.M. 1991. DNA fingerprinting using very short arbitrary oligonucleotides. *Bio/technology*, 9, 553-557.

Caetano-Anoles, G., Bassam, B.J., and Gresshof, P.M. 1993. Enhanced detection of polymorphic DNA by multiple arbitrary amplicon profiling of endonuclease digested DNA; identification of markers tightly linked to the supernodulation locus in soybean. *Mol. Gen. Genet.*, 241, 57-64.

Cagigas, M.E., Vazquez, E., Blanco, G., and Sánchez, J.A. 1999. Combined assessment of genetic variability in populations of brown trout (*Salmo trutta* L.) based on allozymes, microsatellites and RAPD markers. *Mar. Biotechnol.*, 1, 286-296.

Cai, Q., Guy, C.L., and Moore, G.A. 1994. Extension of the linkage map in *Citrus* using random amplified polymorphic DNA (RAPD) markers and RFLP mapping of cold-acclimatation-responsive loci. *Theor. Appl. Genet.*, 89, 606-614.

Camargo, L.E.A., and Osborn, T.C. 1996. Mapping loci controlling flowering time in *Brassica oleracea*. *Theor. Appl. Genet.*, 92, 610-616.

Camefort, H. 1969. Fecondation et proembryogenese chez les Abietacees (notion de neocytoplasme). *Rev. Cytol. et Biol. veg.*, 32, 253-271.

Causse, M.A., Fulton, T.M., Cho, Y.G., Ahn, S.N., Chunwongse, J., Wu, K., Xiao, J., Yu, Z., Ronald, P.C., Harrington, S.E., Second, G., McCouch, S.R., and Tanksley, S.D. 1994. Saturated molecular map of the rice genome based on an interspecific backcross population. *Genetics*, 138, 1251-1274.

Causse, M.A., Rocher, J.P., Henry, A.M., Charcosset, A., Prioul, J.L., and de Vienne, D. 1995. Genetic dissection of the relationship between carbon metabolism and early growth in maize, with emphasis on key-enzyme loci. *Mol. Breeding*, 1, 259-272.

Causse, M., Santoni, S., Damerval, C., Maurice, A., Charcosset, A., Deatrick, J., and de Vienne, D. 1996. A composite map of expressed sequences in maize. *Genome*, 39, 418-432.

Cerna, F.J., Cianzo, S.R., Rafalski, A., Tingey, S., and Dier, D. 1997. Relationship between seed yield heterosis and molecular marker heterozygosity in soybean. *Theor. Appl. Genet.*, 95, 460-467.

Chakraborty, R., and Leimar, O. 1987. Genetic variation within a subdivided population. In *Population Genetics and Fishery Management*, Ryman, N., and Utter, F., eds., University of Washington Press, Seattle and London, pp. 89-120.

Chakravarti, A., Lasher, L.A., and Reefer, J.E. 1991. A maximum likelihood method for estimating genome length using genetic linkage data. *Genetics*, 128, 175-182.

Chalmers, K.J., Barua, U.M., Hackett, C.A., Thomas, W.T.B., Waugh, R., and Powell, W. 1993. Identification of RAPD markers linked to genetic factors controlling the milling energy requirement of barley. *Theor. Appl. Genet.*, 87, 314-320.

Champoux, M.C., Wang, G., Sarkarung, S., Mackill, D.J., O'Toole, J.C., Huang, N., and McCouch, S.R. 1995. Locating genes associated with root morphology and drought avoidance in rice via linkage to molecular markers. *Theor. Appl. Genet.*, 90, 969-981.

Chantereau, J. 1993. Etude de l'heterosis chez le sorgho (*Sorghum bicolor* L. Moench) par l'exploitation d'ecotypes et l'analyse de leurs divergences. 3d cycle thesis, University of Paris-South.

Chao, S., Baysdorfer, C., Heredia-Diaz, O., Musket, T., Xu, G., and Coe, E.H. 1994. RFLP mapping of partially sequenced leaf cDNA clones in maize. *Theor. Appl. Genet.*, 88, 717-721.

Chao, S., Sharp, P.J., and Worland, A.J. 1989. RFLP-based genetic maps of wheat homeologous group 7 chromosomes. *Theor. Appl. Genet.*, 78, 495-504.

Chaparro, J.X., Werner, D.J., O'Malley, D., and Sederoff, R.R. 1994. Targeted mapping and linkage analysis of morphological isozyme, and RAPD markers in peach. *Theor. Appl. Genet.*, 87, 805-815.

Charcosset, A. 1992. Prediction of heterosis. In *Reproductive Biology and Plant Breeding*. Proc. XIIIth Eucarpia meeting, Dattee, Y., Dumas, C., and Gallais, A., eds., Springer-Verlag.

Charcosset, A., Bonnisseau, B., Touchebeuf, O., Burstin, J., Dubreuil, P., Barriere, Y., Gallais, A., and Denis, J.B. 1998. Prediction of maize hybrid silage performance using marker data: comparison of several models for specific combining ability. *Crop Sci.*, 38, 38-43.

Charcosset, A., Dubreuil, P., Essioux, L., and Gallais, A. 1994. Utilisation des techniques de marquage génétique pour l'analyse de la variabilité génétique et la prédiction du phénomène d'hétérosis. *Le Selectionneur Français*, 44, 3-13.

Charcosset, A., and Essioux, L. 1994. The effect of population structure on the relationship between heterosis and heterozygosity at marker loci. *Theor. Appl. Genet.*, 89, 336-343.

Charcosset, A., and Gallais, A. 1996. Estimation of the contribution of quantitative trait loci (QTL) to the variance of a quantitative trait by means of genetic markers. *Theor. Appl. Genet.*, 93, 1193-1201.

Charcosset, A., Lefort-Buson, M., and Gallais, A. 1991. Relationships between heterosis and heterozygosity at marker loci: a theoretical computation. *Theor. Appl. Genet.*, 81, 571-575.

Chen, F.Q., Prehn, D., Hayes, P.M., Mlrooney, D., Corey, A., and Vivar, H. 1994. Mapping genes for resistance to barley stripe rust (*Puccinia striifornua* f. sp. *hordei*). *Theor. Appl. Genet.*, 88, 215-219.

Cheres, M.T., Miller, J.F., Crane, J.M. and Knapp, S.J. 2000. Genetic distance as a predictor of heterosis and hybrid performance within and between heterotic groups in sunflower. *Theor. Appl. Genet.*, 100, 889-894.

Chesnoy, L. 1987. L'origine des organites du cytyplasme embryonnaire chez les gymnospermes. *Bull. Soc. bot. Fr., Actual. Bot.*, 134, 51-56.

Chong, D.K.X., Yang, R.C., and Yeh, F.C. 1994. Nucleotide divergence between populations of trembling aspen (*Populus tremuloides*) estimated with RAPDs. *Cur. Genet.*, 26, 374-376.

Churchill, G.A., and Doerge, R.W. 1994. Empirical threshold values for quantitative trait mapping. *Genetics*, 138, 963-971.

Churchill, G.A., Giovannoni, J.J., and Tanksley, S.D. 1993. Pooled-sampling makes high-resolution mapping practical with DNA markers. *Proc. Natl. Acad. Sci. USA*, 90, 16-20.

Chyi, Y.S., Hoenecke, M.E., and Semyk, J.L. 1992. A genetic map of restriction fragment length polymorphism loci for *Brassica rapa* (syn. *campestris*). *Genome*, 35, 746-757.

Clark, A.G., and Lanigan, C.M.S. 1993. Prospects for estimating nucleotide divergence with RAPDs. *Mol. Biol. Evol.*, 10, 1096-1111.

Concibido, V.C., Young, N.D., Lange, D.A., Denny, R.L., Danesh, D., and Orf, J.H. 1996. Targeted comparative genome analysis and qualitative mapping of a major partial-resistance gene to the soybean cyst nematode. *Theor. Appl. Genet.*, 93, 234-241.

Conner, B.J., Reyes, A.A., Morin, C., Itakura, K., Teplitz, R.L., and Wallace, R.B. 1983. Detection of sickle cell β^S-globin allele by hybridization with synthetic oligonucleotides. *Proc. Natl. Acad. Sci. USA*, 80, 278-282.

Cooke, R., Raynal, M., Laudie, M., Grellet, F., Delseny, M., Morris, P.C., Guerrier, D., Giraudat, J., Quigley, F., Clabault, G., Li, Y.F., Mache, R., Krivitzky, M., Gy, I.J., Kreis, M., Lechanry, A., Parmentier, Y., Marbach, J., Fleck, J., Clement, B., Philipps, G., Herve, C., Bardet, C., Tremousaygue, D., and Hofte, H. 1996. Further progress towards a catalogue of all *Arabidopsis* genes: analysis of a set of 5,000 non-redundant ESTs. *Plant J.*, 9, 101-124.

Costa, P., Pot, D., Dubos, C., Frigerio, J-M., Pionneau, C., Bodénès, C., Bertocchi, E., Cervera, M.T., Remington, D.L., and Plomion, C. 2000. A genetic map of maritime pine based on AFLP, RAPD and protein markers. *Theor. Appl. Genet.*, 100, 39-48.

Crouzillat, D., Lerceteau, E., Petiard, V., Morera, J., Rodriguez, H., Walker, D., Phillips, W., Ronning, C., Schnell, R., Osei, J., and Fritz, P. 1996. *Theobroma cacao* L.: a genetic linkage map and quantitative trait loci analysis. *Theor. Appl. Genet.*, 93, 205-214.

Crow, J.F. 1990. Mapping functions. *Genetics*, 125, 669-671.

Crow, J.F., and Kimura, M. 1964. The number of alleles that can be maintained in a finite population. *Genetics*, 49, 725-738.

Culley, T.M., and Wolfe, D. 2001. Population genetic structure of the cleistogamous plant species *Viola pubescens* Aiton (Violaceae), as indicated by allozyme and ISSR molecular markers. *Heredity*, 86, 545-556.

Da Silva, J.A.G., Sorrels, M.E., Burnquist, W.L., and Tanksley, S.D. 1993. RFLP linkage map and genome analysis of *Saccharum spontaneum*. *Genome*, 36, 782-791.

Dallas, J.F. 1988. Detection of DNA "fingerprintings" of cultivated rice by hybridization with a human microsatellite DNA probe. *Proc. Natl. Acad. Sci. USA*, 85, 6831-6835.

Damerval, C., and de Vienne, D. 1985. Divergence morphologique et divergencemoleculaire: 1. Apport des marqueurs proteiques. In *Les Distances Genetiques*, Lefort-Buson, M., and de Vienne, D., eds., INRA, Paris.

Damerval, C., Maurice, A., Josse, J.M., and de Vienne, D. 1994. Quantitative trait loci underlying gene product variation—a novel perspective for analyzing regulation of genome expression. *Genetics*, 137, 289-301.

Darvasi, A., and Soller, M. 1992. Selective genotyping for determination of linkage between a marker locus and a quantitative trait locus. *Theor. Appl. Genet.*, 85, 353-359.

Darvasi, A., and Soller, M. 1994. Selective DNA pooling for a determination of linkage between a molecular marker and a quantitative trait locus. *Genetics*, 138, 1365-1373.

Darvasi, A., Weinreb, A., Minke, V., Weller, J.I., and Soller, M. 1993. Detection marker-QTL linkage and estimating QTL gene effect and map location using a saturated genetic map. *Genetics*, 134, 943-951.

De Vicente, M.C., and Tanksley, S.D. 1991. Genome-wide reduction in recombination of backcross progeny derived from male versus female gametes in an interspecific cross of tomato. *Theor. Appl. Genet.*, 83, 173-178.

De Vicente, M.C., and Tanksley, S.D. 1993. QTL analysis of transgressive segregation in an interspecific tomato cross. *Genetics*, 134, 585-596.

de Vienne, D. 1990. L'analyse du determinisme des caracteres quantitatifs chez les vegetaux. *Medecine/Sciences*, 6, XI-XVI

de Vienne, D. 1995. Strategies de caracterisation des locus a effets quantitatifs. *Le Selectionneur Francais*, 46, 19-25.

de Vienne, D., Burstin, J., Gerber, S., Leonardi, A., Le Guilloux, M., Murigneux, A., Beckert, M., Bahrman, N., Damerval, C., and Zivy, M. 1996. Two-dimensional electrophoresis of proteins as a source of monogenic and codominant markers for population genetics and mapping the expressed genome. *Heredity*, 76, 166-177.

de Vienne, D., and Damerval, C. 1985. Mesures de la divergence genetique: 3. Distances calculees a partir de marqueurs moleculaires. In *Les Distances Genetiques*, Lefort-Buson, M., and de Vienne, D., eds., INRA, Paris.

De Wolf, H., Backeljau, T., and Verhagen, R. 1998. Congruence between allozyme and RAPD data in assessing macrogeographical genetic variation in the periwinkle *Littorina striata* (Mollusca, Gastropoda). *Heredity*, 81, 486-492.

Dekkers, J.C., and Hospital, F. 2002. The use of molecular genetics in the improvement of agricultural populations. *Nat. Rev. Genet.*, 3, 22-32.

Delourme, R., Bouchereau, A., Hubert, N., Renard, M., and Landry, B.S. 1994. Identification of RAPD markers linked to a fertility restorer gene for the Ogura radish cytoplasmic male sterility of rapeseed (*Brassica napus* L.). *Theor. Appl. Genet.*, 88, 741-748.

Demesure, B., Sodzi, N., and Petit, R.J. 1995. A set of universal primers for amplification of polymorphic non-coding regions of mitochondrial and chloroplast DNA in plants. *Mol. Ecol.*, 4, 129-131.

Desplanque, B., Boudry, P., Broomberg, K., Saumitou-Laprade, P., Cuguen, J., Van Dijk, H. 1999. Genetic diversity and gene flow between wild, cultivated and weedy forms of *Beta vulgaris* L. (Chenopodiaceae), assessed by RFLP and microsatellite markers. *Theor. Appl. Genet.*, 98, 1194-1201.

Deu, M., Gonzalez de Leon, D., Glaszmann, J.C., Degremont, I., Chantereau, J., Lanaud, C., and Hamon, P. 1994. RFLP diversity in cultivated sorghum in relation to racial differentiation. *Theor. Appl. Genet.*, 88, 838-844.

Devey, M.E., Delfino-Mix, A., Kinloch, B.B., and Neale, D.B. 1995. Random polymorphic DNA markers tightly linked to a gene for resistance to white pine blister rust in sugar pine. *Proc. Natl. Acad. Sci. USA*, 92, 2066-2070.

Devey, M.E., Fiddler, T.A., Lie, B.H., Knapp, S.J., and Neale, B.D. 1994. An RFLP linkage map for loblolly pine based on a three-generation outbred pedigree. *Theor. Appl. Genet.*, 99, 273-278.

Devos, K.M., Atkinson, M.D., Chinoy, C.N., Francis, H.A., Harcourt, R.L., Koebner, R.M.D., Liu, C.J., Masoje, P., Xie, D.X, and Gale, M.D. 1993. Chromosomal rearrangements in the rye genome relative to that of wheat. *Theor. Appl. Genet.*, 85, 673-680.

Devos, K.M., Atkinson, M.D., Chinoy, C.N., Liu, C.J., and Gale, M.D. 1992. RFLP-based genetic map of the homoeologous group 3 chromosomes of wheat and rye. *Theor. Appl. Genet.*, 83, 931-939.

Devos, K.M., and Gale, M.D. 1997. Comparative genetics in the grasses. *Plant Mol. Biol.*, 5, 3-15.

Devos, K.M., and Gale, M.D. 2000. Genome relationships: the grass model in current research. Plant Cell., 12, 637-646.

Díaz, O., Sun, G.L., Salomon, B., and von Bothmer, R. 2000. Levels and distribution of allozyme and RAPD variation in populations of *Elymus fibrosus* (Schrenk) Tzvel. (Poaceae). *Genet. Resources Crop Evol.*, 47, 11-24.

Diers, B.W., MacVetty, P.B.E. and Osborn, T.C. 1996. Relationship between heterosis and genetic distance based on restriction length polymorphism markers in oilseed rape (*Brasica napus* L.). *Crop Sci.*, 36, 79-83.

Diers, B.W., Mansur, L., Imsande, J., and Shoemaker, R.C. 1992. Mapping *Phytophthora* resistance loci in soybean with restriction fragment length polymorphism markers. *Crop Sci.*, 32, 377-383.

Dillmann, C., Bar-Hen, A., Guerin, D., Charcosset, A., and Murigneux, A. 1997a. Comparison of RFLP and morphological distances between maize *Zea mays* L. inbred lines. Consequences for germplasm protection purposes. *Theor. Appl. Genet.*, 95, 92-102.

Dillmann, C., Charcosset, A., Goffinet, B., Smith, J.S.C., and Dattee, Y. 1997b. Best linear unbiased estimators of the molecular genetic distance between inbred lines. In *Advances in Biometrical Genetics*, Krajewski, P., and Kaczmarek, Z., eds., Poznan, pp. 105-110.

Dirlewanger, E., Isaac, P.G., Ranade, S., Belajouza, M., Cousin, R., and de Vienne, D. 1994. Restriction fragment length polymorphism analysis of loci associated with disease resistance genes and developmental traits in *Pisum sativum* L. *Theor. Appl. Genet.*, 88, 17-27.

Doebley, J., and Stec, A. 1991. Genetic analysis of the morphological differences between maize and teosinte. *Genetics*, 129, 285-295.

Doebley, J., and Stec, A. 1993. Inheritance of the morphological differences between maize and teosinte: comparison of results for two F_2 populations. *Genetics*, 134, 559-570.

Doebley, J., Stec, A., and Gustus, C. 1995. *Teosinte branched1* and the origin of maize: evidence for epistasis and the evolution of dominance. *Genetics*, 141, 333-346.

Doebley, J., Stec, A., and Hubbard, L. 1997. The evolution of apical dominance in maize. *Nature*, 386, 485-488.

Dong, J., and Wagner, D.B. 1994. Paternally inherited chloroplast polymorphism in *Pinus*: estimation of diversity and population subdivision, and tests of disequilibrium with a maternally inherited mitochondrial polymorphism. *Genetics*, 136, 1187-1194.

Donner, H.K. 1986. Genetic fine structure of the bronze locus in maize. *Genetics*, 113, 1021-1036.

Dow, B.D., and Ashley, M.V. 1998. High levels of gene flow in bur oak revealed by paternity analysis using microsatellites. *J. Hered.*, 89, 62-70.

Dow, B.D., and Ashley, M.V. 1996. Microsatellite analysis of seed dispersal and parentage of saplings in bur oak, *Quercus macrocarpa. Mol. Ecol.*, 5, 615-627.

Dubreuil, P., and Charcosset, A. 1998. Genetic diversity within and among maize populations: a comparison between isozymes and nuclear RFLP loci. *Theor. Appl. Genet.*, 96, 577-587.

Dubreuil, P., Dufour, P., Krejci, E., Causse, M., de Vienne, D., Gallais, A., and Charcosset, A. 1996. Organization of RFLP diversity among inbred lines of maize representing the most significant heterotic groups. *Crop Sci.*, 36, 790-799.

Dubreuil P., Charcosset A. (1998). Genetic diversity within and among maize populations: a comparison between isozyme and nuclear RFLP loci. *Theor. Appl. Genet.*, 96, 577-587.

Dudley, J.W., Saghai Maroof, M.A., and Rufener, G.K. 1991. Molecular markers and grouping of parents in maize breeding programs. *Crop Sci.*, 31, 718-723.

Dumolin-Lapegue, S., Piemonge, M.H., and Petit, R.J. 1996. An enlarged set of consensus primers for the study of organelle DNA in plants. *Mol. Ecol.*, 6, 393-397.

Durham, R.E., Liou, P.C., Gmitter, F.G., and Moore, G.A. 1992. Linkage of restriction fragment length polymorphisms and isozymes in *Citrus*. *Theor. Appl. Genet.*, 84, 39-48.

East, E.M. 1915. Studies on size inheritance in *Nicotiana. Genetics*, 1, 164-176.

Echt, C.S., Kidwell, K.K., Knapp, S.J., Osborn, T.C., and McCoy, T.J. 1994. Linkage mapping in diploid alfalfa (*Medicago sativa*). *Genome*, 37, 61-71.

Ecke, W., Uzunova, M., and Weissleder, K. 1995. Mapping the genome of rapeseed (*Brassica napus* L.) II. Localization of genes controlling erucic acid synthesis and seed oil content. *Theor. Appl. Genet.*, 91, 972-978.

Edwards, K.J., Barker, J.H.A., Daly, A., Jones, C., and Karp, A. 1996. Microsatellite libraries enriched for several microsatellite sequences in plants. *Biotechniques*, 20, 758-760.

Edwards, M.D., Helentjaris, T., Wright, S., and Stuber, C.W. 1992. Molecular-marker facilitated investigations of QTL in maize. IV. Analysis based on genome saturation and restriction fragment length polymorphism markers. *Theor. Appl. Genet.*, 83, 765-774.

Edwards, M.D., Stuber, C.W., and Wendel, J.F. 1987. "Molecular-marker facilitated investigations of QTL in maize. I. Number, genomic distribution and types of gene action. *Genetics*, 116, 113-125.

El-Din El-Assal, S., Alonso-Blanco, C., Peters, A.J., Raz, V., and Koornneef, M. 2001. A QTL for flowering time in *Arabidopsis* reveals a novel allele of CRY2. *Nat Genet.*, 29, 435-440.

El Mousadik, A., and Petit, R.J. 1996a. High levels of genetic differentiation for allelic richness among populations of the argan tree (*Argania spinosa* (L.) Skeels) endemic to Morocco. *Theor. Appl. Genet.*, 92, 832-839.

El Mousadik, A., and Petit, R.J. 1996b. Chloroplast DNA phylogeography of the argan tree of Morocco. *Mol. Ecol.*, 5, 547-555.

Ellis, T.H.N., Turner, L., Hellens, R.P., Lee, D., Harker, C.L., Enard, C., Domoney, C., and Davies, D.R. 1992. Linkage maps in pea. *Genetics*, 130, 649-663.

Ellstrand, N.C. 1992. Gene flow among seed plant populations. *New Forests*, 6, 241-256.

Emerson, R.A., Beadle, G.W., and Fraser, A.C. 1935. A summary of linkage studies in maize. *Cornell Univ. Agr. Exp. Stn. Mem.*, pp. 1-80.

Ennos, R.A. 1994. Estimating the relative rates of pollen and seed migration among plant populations. *Heredity*, 72, 250-259.

Eshed, Y., Abu-Abied, M., Saranga, Y., and Zamir, D. 1992. *Lycopersicon esculentum* lines containing small overlapping introgressions from *L. pennellii*. *Theor. Appl. Genet.*, 83, 1027-1034.

Eshed, Y., and Zamir, D. 1995. An introgression line population of *Lycopersicon pennellii* in the cultivated tomato enables the identification and fine mapping of yield associated QTLs. *Genetics*, 141, 1147-1162.

Estoup, A., Rousset, F., Michalakis, Y., Cornuet, J.M., Adriamanga, M., and Guyomard, R. 1998. Comparative analysis of microsatellite and allozyme markers: a case study investigating microgeographic differentiation in brown trout (*Salmo trutta*). *Mol. Ecol.*, 7, 339-353.

Excoffier, L., Smouse, P.E., and Quattro, J.M. 1992. Analysis of molecular variance inferred from metric distances among DNA haplotypes: application to human mitochondrial DNA restriction data. *Genetics*, 131, 479-491.

Falconer, D.S. 1961. *Introduction to Quantitative Genetics*, 2nd ed. Longman, London.

Falconer, D.S., and MacKay, T.F.C. 1996. Quantitative trait loci. In *Introduction to Quantitative Genetics*, 4th ed., Longman Group Ltd., Harlow, England.

Fatokun, C.A., Menancio-Hautea, D.I., Danesh, D., and Young, N.D. 1992. Evidence for orthologous seed weight genes in cowpea and mung bean based on RFLP mapping. *Genetics*, 132, 841-846.

Faure, S., Noyer, J.L., Horry, J.P., Bakry, F., Lanaud, C., and Gonzalez de Leon, D. 1994. A molecular marker-based linkage map of diploid bananas (*Musa acuminata*). *Theor. Appl. Genet.*, 87, 517-526.

Ferguson, M.E., Newbury, H.J., Maxted, N., Ford-Lloyd, B.V., and Robertson, L.D. 1998. Population genetic structure in *Lens* taxa revealed by isozyme and RAPD analysis. *Genet. Resources Crop Evol.*, 45, 549-559.

Ferreira, M.E., Williams, P.H., and Osborn, T.C. 1994. RFLP mapping of *Brassica napus* using doubled haploid lines. *Theor. Appl. Genet.*, 89, 615-621.

Fisher, R.A. 1937. *The Design of Experiments*, 2nd ed. Oliver & Boyd, Edinburgh, London, 260 pp.

Flavell, R.B. 1985. Chromosome architecture: the distribution of recombination sites, the structure of ribosomal DNA loci and the multiplicity of sequences containing inverted repeats. In *Molecular Form and Function of the Plant Genome*, van Vloten-Dotin, L., Groot, G.S.P., and Hall, I.C., eds., NATO ASI, 83, Plenum Press, New York, pp. 1-14.

Frary, A., Nesbitt, T.C., Grandillo, S., Knaap, E., Cong, B., Liu, J., Meller, J., Elber, R., Alpert, K.B., and Tanksley, S.D. 2000. *fw2.2*: a quantitative trait locus key to the evolution of tomato fruit size. *Science*, 289, 85-88.

Freeling, M. 2001. Grasses as a single genetic system: reassessment 2001. *Plant Physiol.*, 125, 1191-1197.

Frei, O.M., Stuber, C.W., and Goodman, M.M. 1986. Use of allozymes as genetic markers for predicting performance in maize single cross hybrids. *Crop Sci.*, 26, 37-42.

Freymark, P.J., Lee, M., Woodman, W.L., and Martinson, C.A. 1993. Quantitative and qualitative trait loci affecting host-plant response to *Exserohilum turcicum* in maize (*Zea mays* L.). *Theor. Appl. Genet.*, 87, 537-544.

Freyre, R., Warnke, S., Sosinski, B., and Douches, D.S. 1994. Quantitative trait locus analysis of tuber dormancy in diploid potato (*Solanum* spp.). *Theor. Appl. Genet.*, 89, 474-480.

Fridman, E., Pleban, T., and Zamir D. 2000. A recombinant hotspot delimits a wild QTL for tomato sugar content to 484 bp within an invertase gene. *Proc. Natl. Acad. Sci. USA*, 97, 4718-4723.

Frova, C., and Sari-Gorla, M.S. 1993. Quantitative maize expression of HSPs: genetic dissection and association with thermotolerance. *Theor. Appl. Genet.*, 86, 213-220.

Fuerst, P.A., and Ferrell, R.E. 1980. The stepwise mutation model: an experimental evaluation utilizing hemoglobin variants. *Genetics*, 94, 185-201.

Fukuoka, S., Inoue, T., Miyao, A., Monna, L., Zhong, H.S., Sasaki, T., and Minobe, Y. 1994. Mapping of sequence-tagged sites in rice by single-strand conformation polymorphism. *DNA Res.*, 1, 271-277.

Gallais, A. 1990. *Theorie de la Selection en Amelioration des Plantes*. Masson ed. 588 p.

Gallais, A. 1995. La selection assistee par marqueurs. *Le Seletionneur Francais*, 43, 43-62.

Gallais, A., and Charcosset, A. 1994. Efficiency of marker-assisted selection. In *Biometrics in Plant Breeding: Applications of Molecular Markers*. Proc. 9th meeting of the Eucarpia Section Biometrics in Plant Breeding, van Oijen, J.W., and Jansen, J., eds., pp. 91-99.

Gallais A., Dillmann C., and Hospital F. 1997. An analytical approach of marker-assisted selection with selection on markers only. *In Proc. 10th Meeting of the Eucarpia Section Biometrics in Plant Breeding*, Krajewski, P., and Kaczmarek, Z., eds., IGR, Poznan, pp. 111-116.

Gallais, A., and Rives, M. 1993. Detection, number and effects of QTLs for a complex character. *Agronomie*, 13, 723-738.

Gardiner, J.M., Coe, E.H., Melia-Hancock, S., Hoisington, D.A., and Chao, S. 1993. Developpment of a core RFLP map using an immortalized F$_2$ population. *Genetics*, 134, 917-930.

Gardner, C.O., and Eberhart, S.A. 1966. Analysis and interpretation of the variety cross diallele and related populations. *Biometrics*, 22, 439-452.

Gebhardt, C., Ritter, E., Barone, A., Debener, T., Walkemeier, B., Schachts-chabel, U., Kaufmann, H., Thompson, R.D., Bonierbale, M.W., Ganal, M.W., Tanksley, S.D., and Salamini, F. 1991. RFLP maps of potato and their alignment with the homeologous tomato genome. *Theor. Appl. Genet.*, 83, 49-57.

Gebhardt, C., Ritter, E., Debener, T., Schachtschabel, U., Walkemeier, B., Uhrig, H., and Salamini, F. 1989. RFLP analysis and linkage mapping in *Solanum tuberosum*. *Theor. Appl. Genet.*, 78, 65-75.

Gentzbittel, L., Zhang, Y.X., Vear, F., Griveau, B., and Nicolas, P. 1994. RFLP studies of genetic relationships among inbred lines of the cultivated sunflower, *Helianthus annus* L.: evidence for distinct restorer and maintainer germplasm pools. *Theor. Appl. Genet.*, 89, 419-425.

Gerber, S., Mariette, S., Streiff, R., Bodénès, C., and Kremer A. 2000. Comparison of microsatellites and amplified fragment length polymorphism markers for parentage analysis. *Mol. Ecol.*, 9, 1037-1048.

Gerber, S., and Rodolphe, F. 1994a. Estimation and test for linkage between markers: a comparison of LOD score and χ^2 test in a linkage study of maritime pine (*Pinus pinaster* Ait.). *Theor. Appl. Genet.*, 88, 293-297.

Gerber, S., and Rodolphe, F. 1994b. An estimation of the genome length of maritime pine (*Pinus pinaster* Ait.). *Theor. Appl. Genet.*, 88, 289-292.

Gerber, S., Rodolphe, F., Bahrman, N., and Baradat, P.H. 1993. Seed-protein variation in matirime pine (*Pinus pinaster* Ait.) revealed by two-dimensional electrophoresis: genetic determinism and construction of a linkage map. *Theor. Appl. Genet.*, 85, 521-528.

Ghareyazie, B., Huang, N., Second, G., Bennet, J., and Kush, G.S. 1995. Classification of rice germplasm. I. Analysis using ALP and PCR-based RFLP. *Theor. Appl. Genet.*, 91, 218-227.

Gimelfarb, A., and Lande, R. 1994. Simulation of marker-assisted selection in hybrid populations. *Genet. Res. Camb.*, 63, 39-47.

Gimelfarb, A., and Lande, R. 1995. Marker-assisted selection and marker-QTL associations in hybrid populations. *Theor. Appl. Genet.*, 91, 522-528.

Giordano, M., Oefner, P.J., Underhill, P.A., Cavalli Sforza, L.L., Tosi, R., and Richiardi P.M. 1999. Identification by denaturing high-performance liquid chromatography of numerous polymorphisms in a candidate region for multiple sclerosis susceptibility. *Genomics*, 15, 247-253.

Giraudat, J., Hauge, B.H., Valon, C., Smalle, J., Parcy, F., and Goodman, H.M. 1992. Isolation of the *Arabidopsis ABI3* gene by positional cloning. *The Plant Cell*, 4, 1251-1261.

Godoy, J.A., and Jordano, P. 2001. Seed dispersal by animals: exact identification of source trees with endocarp DNA microsatellites. *Mol. Ecol.*, 10, 2275-2283.

Godshalk, E.B., Lee, M., and Lamkey, K.R. 1990. Relationship of restriction fragment length polymorphisms to single-cross hybrid performance of maize. *Theor. Appl. Genet.*, 80, 273-280.

Godwin, I.D., Aitken, E.A., Smith, and L.W. 1997. Application of inter simple sequence repeat (ISSR) marker to plant genetics. *Electrophoresis*, 18(9): 1524-1528.

Goldman, I.L., Rocheford, T.R., and Dudley, J.W. 1993. Quantitative trait loci influencing protein and starch concentration in the Illinois Long Term Selection maize strains. *Theor. Appl. Genet.*, 87, 217-224.

Goldman, I.L., Rocheford, T.R., and Dudley, J.W. 1994. Molecular markers associated with maize kernels oil concentrations in an Illinois high protein × Illinois low protein cross. *Crop Sci.*, 34, 908-915.

Goldman, I.L., Paran, I., and Zamir, D. 1995. Quantitative trait locus analysis of a recombination inbred line population derived from a *Lycopersicon esculentum* × *Lycopersicon cheesmanii* cross. *Theor. Appl. Genet.*, 90, 925-932.

Gonzales-Candelas, F., Elena, S.F., and Moya, A. 1995. Approximate variance of nucleotide divergence between two sequences estimated from restriction fragment data. *Genetics*, 140, 1443-1446.

Gouesnard, B., Bataillon, T.M., Decoux, G., Rozale, C., Schoen, D.J., and David, J.L. 2001. MSTRAT: An algorithm for building germplasm core collections by maximizing allelic or phenotypic richness. *J. Heredity*, 92, 93-94.

Grandillo, S., and Tanksley, S.D. 1996. QTL analysis of horticultural traits differentiating the cultivated tomato from the closely related species *Lycopersicon pimpinellifolium*. *Theor. Appl. Genet.*, 92, 935-951.

Graner, A., Jahorr, A., Schondelmaier, J., Siedler, H., Pilee'n, K., Fishbeck, G., Wenzel, G., and Herrmann, R.G. 1991. Construction of an RFLP map of barley. *Theor. Appl. Genet.*, 83, 250-256.

Grant D., Cregan P., Shoemaker R.C. Genome organization in dicots: genome duplication in Arabidopsis and synteny between soybean and Arabidopsis. 2000. *Proc. Natl. Acad. Sci. USA.*, 97, 4168-4173.

Grattapaglia, D., Bertolucci, F.L., and Sederoff, R.R. 1995. Genetic mapping of QTLs controlling vegetative propagation in *Eucalyptus grandis* and *E. urophylla* using a pseudo-testcross mapping strategy and RAPD markers. *Theor. Appl. Genet.*, 90, 933-947.

Grattapaglia, D., and Sederoff, R. 1994. Genetic linkage maps of *Eucalyptus grandis* and *E. urophylla* using a pseudo-testcross mapping strategy and RAPD markers. *Genetics*, 137, 1121-1137.

Gregorius, H.R. 1980. The probability for losing an allele when diploid genotypes are sampled. *Biometrics*, 36, 643-652.

Grivet, D., Heinze, B., Vendramin, G.G., and Petit, R.J. 2002. Genome walking with consensus primers: application to the large single copy region of chloroplast DNA. *Mol. Ecol.* notes (in press).

Grivet, L., D'Hont, A., Dufour, P., Hamon, P., Roques, D., and Glaszmann, J.C. 1994. Comparative genome mapping of sugar cane with other species within the Andropogoneae tribe. *Heredity*, 73, 500-508.

Hackett, C.A., and Weller, J.I. 1995. Genetic mapping of quantitative trait loci for traits with ordinal distributions. *Biometrics*, 51, 1252-1263.

Hahn, M., Serth, J., Fislage, R., Wolfes, H., Allhoff, E., Jonas, V., and Pingoud, A. 1993. Polymerase chain reaction detection of a highly polymorphic VNTR segment in intron 1 of the human p53 gene. *Clin. chem.*, 39, 549-550.

Haldane, J.B.S. 1919. The combination of linkage values and the calculation of distance between the loci of linked factors. *J. Genet.*, 8, 299-309.

Haldane, J.B.S., and Waddington, C. 1931. Inbreeding and linkage. *Genetics*, 16, 357-374.

Haley, C.S., and Knott, S.A. 1992. A simple regression method for mapping quantitative trait loci in line crosses using flanking markers. *Heredity*, 69, 315-324.

Haley, S.D., Mikals, P.N., Stavely, J.R., Byrum, J., and Kelly, J.D. 1993. Identification of RAPD markers linked to a major rust resistance gene block in common bean. *Theor. Appl. Genet.*, 86, 505-512.

Halward, T., Stalker, H.T., and Kochert, G. 1993. Development of an RFLP linkage map in diploid peanut species. *Theor. Appl. Genet.*, 87, 379-384.

Hammond-Kosack, K.E., and Jones J.D.G. 1997. Plant disease resistance genes. *Annu. Rev. Plant Physiol. Plant Mol. Biol.*, 48, 575-607.

Hamrick, J.L., and Godt, M.J.W. 1989. Allozyme diversity in plant species. In *Plant Population Genetics, Breeding and Genetic Resources*, Brown, A.H.D., Clegg, M.T., Kahler, A.L., and Weir, B.S., eds. Sinauer Associates, pp. 43-64.

Han, F., Ullrich, S.E., Clancy, J.A., Jitkov, V., Kilian, A., and Romagosa, I. 1996. Verification of barley seed dormancy loci via linked molecular markers. *Theor. Appl. Genet.*, 92, 87-91.

Hanson, W.D. 1959. Early generation analysis of length of heterozygous chromosome segments around a locus held heterozygous with backcrossing or selfing. *Genetics*, 44, 833-837.

Harris, S.A., and Ingram, R. 1991. Chloroplast DNA and biosystematics: the effect of intraspecific diversity and plastid transmission. *Taxon*, 40, 393-412.

Hartl, D.L., and Clark, A.D. 1989. *Principles of Population Genetics*. Sinauer Associates, 682 pp.

Havey, M.J., and Muehlbauer, F.J. 1989. Linkages between restriction fragment length, isozyme and morphological markers in lentil. *Theor. Appl. Genet.*, 77, 395-401.

Hayes, H.K., and Johnson, I.J. 1939. The breeding of improved selfed lines of corn. *J. Am. Soc. Agron.*, 31, 710-724.

Hayes, P.M., Lui, B.H., Knapp, S.J., Chen, F., Jones, B., Blake, T., Franckowiak, J., Rasmusson, D., Sorrells, M., Ullrichs, E., Wesenberg, D., and Kleinhofs, A. 1993. Quantitative trait locus effects and environmental interaction in a sample of North American barley germplasm. *Theor. Appl. Genet.*, 87, 392-401.

Heid, C.A., Stevens, J., Livak, K.J., and Williams, P.M. 1996. Real time quantitative PCR. *Genome Res.*, 6, 986-994.

Heidrich-Sobrinho, E., and Cordeiro, A.R. 1975. Codominant isoenzymic alleles as markers of genetic diversity correlated with heterosis in maize (*Zea mays*). *Theor. Appl. Genet.*, 46, 197-199.

Helentjaris, T. 1987. A linkage map for maize based on RFLPs. *Trends Genet.*, 3, 217-221.

Helentjaris, T., Weber, D.F., and Wright, S. 1986a. Use of monosomics to map cloned DNA fragments in maize. *Proc. Natl. Acad. Sci. USA*, 83, 6035-6039.

Helentjaris, T., Slocum, M., Wright, S., Schaeffer, A., and Nienhuis, J. 1986b. Construction of genetic linkage maps in maize and tomato using restriction fragment length polymorphisms. *Theor. Appl. Genet.*, 72, 761-769.

Helentjaris, T., Weber, D., and Wright, S. 1988. Identification of the genomic locations of duplicate nucleotide sequences in maize by analysis of restriction fragment length polymorphisms. *Genetics*, 118, 353-363.

Henderson, C.R. 1975. Best linear unbiased estimation and prediction under a selection model. *Biometrics*, 31, 423-447.

Heun, M. 1992. Mapping quantitative powdery mildew resistance of barley using a restriction fragment length polymorphism map. *Genome*, 35, 1019-1025.

Heun, M., Kennedy, A.E., Anderson, J.A., Lapitan, N.L.V., Sorrells, M.E., and Tanksley, S.D. 1991. Construction of a restriction fragment length polymorphism map for barley (*Hordeum vulgare*). *Genome*, 34, 437-444.

Hinze, K., Thompson, R.D., Ritter, E., Salamini, F., and Schulze-Lefert, P. 1991. Restriction fragment length polymorphism-mediated targeting of the *ml-o* resistance locus in barley (*Hordeum vulgare*). *Proc. Natl. Acad. Sci. USA*, 88, 3691-3695.

Hipkins, V.D., Krutovskii, K.V., and Strauss, S.H. 1994. Organelle genomes in conifers: Structure, evolution and diversity. *Forest Genet.*, 1(4), 179-189.

Holsinger, K.E., and Mason-Gamer, R.J. 1996. Hierarchical analysis of nucleotide diversity in geographically structured populations. *Genetics*, 142, 629-639.

Hormaza, J.I., Dollo, L., and Polito, V.S. 1994. Identification of a RAPD marker linked to sex determination in *Pistacia vera* using bulked segregant analysis. *Theor. Appl. Genet.*, 89, 9-13.

Hospital, F., and Charcosset, A. 1997. Marker-assisted introgression of quantitative trait loci. *Genetics*, 147, 1469-1485.

Hospital, F., Chevalet, C., and Mulsant, P. 1992. Using markers in gene introgression breeding programs. *Genetics*, 132, 1199-1210.

Hospital, F., Goldringer, I., and Openshaw, S. 2000. Efficient marker based recurrent selection for multiple quantitative trait loci. *Genet. Res. Camb.*, 75, 357-368.

Hospital, F., Moreau, L., Lacoudre, F., Charcosset, A., and Gallais, A. 1997. More on the efficiency of marker-assisted selection. *Theor. Appl. Genet.*, 95, 1181-1189.

Howell, W.M., Jobs, M., Gyllensten, U., and Brooks, A.J. 1999. Dynamic allele-specific hybridization: a new method for scoring single nucleotide polymorphisms. *Nat. Biotechnol.*, 17(1), 87-88.

Huang, B.X., Peakall, R., and Hanna, P.J. 2000. Analysis of genetic structure of blacklip abalone (*Haliotis rubra*) populations using RAPD, minisatellite and microsatellite markers. *Mar. Biol.*, 136, 207-216.

Huang, H., Dane, F., and Kubisiak, T.L. 1998. Allozyme and RAPD analysis of the genetic diversity and geographic variation in wild populations of the American chestnut (Fagaceae). *Amer. J. Bot.*, 85, 1013-1021.

Hudson, R.R. 1990. Gene genealogies and the coalescent process. *Oxf. Surv. Evol. Biol.*, 7, 1-44.

Hudson, R.R., Slatkin, M., and Maddison, W.P. 1992. Estimation of levels of gene flow from DNA sequence data. *Genetics*, 132, 583-589.

Hulbert, S.H., Hott, T.W., Legg, E.J., Lincoln, S.E., Lander, E.S., and Michelmore, R.W. 1988. Genetic analysis of the fungus, *Bremia lactucae*, using restriction length polymorphism. *Genetics*, 120, 947-958.

Hulbert, S.H., Richter, T.E., Axtell, J.D., and Bennetzen, J.L. 1990. Genetic mapping and characterization of sorghum and related crops by means of maize DNA probes. *Proc. Natl. Acad. Sci. USA*, 87, 4251-4255.

Hunter, R.B., and Kannenberg, L.W. 1971. Isozyme characterisation of corn (*Zea mays*) inbreds and its relationship to single cross hybrid performance. *Can. J. Genet. Cytol.*, 13, 649-655.

Hurlbert, S.H. 1971. The nonconcept of species diversity: a critique and alternative parameters. *Ecology*, 52, 577-586.

Iannone, M.A., Taylor, J.D., Chen, J., Li, M.S., Rivers, P., Slentz-Kesler, K.A., and Weiner, M.P. 2000. Multiplexed single nucleotide polymorphism genotyping by oligonucleotide ligation and flow cytometry. *Cytometry*, 39, 131-140.

Innan, H., Terauchi, R., Kahl, G., and Tajima, F. 1999. A method for estimating nucleotide diversity from AFLP data. *Genetics*, 151, 1157-1164.

Inoue, T., Zhong, H.S., Miyao, A., Ashikawa, I., Monna, L., Fukuoka, S., Miyadera, N., Nagamura, Y., Kurata, N., Sasaki, T., and Minobe, Y. 1994. Sequence-tagged sites (STSs) as landmarks in the rice genome. *Theor. Appl. Genet.*, 89, 728-734.

Isabel, N., Beaulieu, J., and Bousquet, J. 1995. Complete congruence between gene diversity estimates derived from genotypic data at enzymes and random amplified polymorphic DNA loci in black spruce. *Proc. Natl. Acad. Sci. USA*, 92, 6369-6373.

Jain, A., Bhatia, S., Banga, S.S., Prakash, S., and Lakshmikumaran, M. 1994. Potential use of random amplified polymorphic DNA (RAPD) technique to study the genetic diversity in Indian mustard (*Brassica juncea*) and its relationship to heterosis. *Theor. Appl. Genet.*, 88, 116-122.

Jansen, R.C. 1993. Interval mapping of multiple quantitative trait loci. *Genetics*, 135, 205-211.

Jansen, R.C., and Stam, P. 1994. High resolution of quantitative traits into multiple loci via interval mapping. *Genetics*, 136, 1447-1455.

Jansen, R.C., van Ooijen, J.W., Stam, P., Lister, C., and Dean, C. 1995. Genotype-by-environment interaction in genetic mapping of multiple quantitative trait loci. *Theor. Appl. Genet.*, 91, 33-37.

Jarne, P., and Lagoda, P.J.L. 1996. Microsatellites, from molecules to populations and back. *Tree*, 11, 424-429.

Jarrel, D.C., Roose, M.L., Traugh, S.N., and Kupper, R.S. 1992. A genetic map of citrus based on the segregation of isozymes and RFLPs in an intergeneric cross. *Theor. Appl. Genet.*, 84, 49-56.

Jeffreys, A.J., Wilson, V., and Thein, S.L. 1985. Hypervariable minisatellite regions in human DNA. *Nature*, 314, 67-73.

Jeunemaitre, X., Soubrier, F., Kotelevtsev, Y.V., Lifton, R.P., Williams, R.R., Lalouel, J.M., and Corvol, P. 1992. Molecular basis of human hypertension: Role of angiotensinogen. *Cell*, 71, 169-180.

Jones, E.S., Liu, C.J., Gale, M.D., Hash, C.T., and Witcombe, J.R. 1995. Mapping quantitative trait loci for downy mildew resistance in pearl millet. *Theor. Appl. Genet.*, 91, 448-457.

Jones, J.D.G. 2001. Putting knowledge of plant disease resistance genes to work. *Curr. Opin. Plant Biol.*, 4, 281-287.

Jukes, T.H., and Cantor, C.R. 1969. Evolution of protein molecules. In *Mammalian Protein Metabolism*, H.N. Munro, ed., Academic Press, pp. 21-132.

Kaback, D.B., Guacci, V., Barber, D., and Mahon, J.W. 1992. Chromosome size dependent control of meiotic recombination. *Science*, 256, 228-232.

Kacser, H., and Burns, J.A. 1981. The molecular basis of dominance. *Genetics*, 97, 639-666.

Kaplan, N. 1983. Statistical analysis of restriction map data and nucleotide sequence data. In *Statistical Analysis of DNA Sequence Data*, B.S. Weir, ed., Mark Dekker, pp. 75-106.

Karl, S.A., and Avise, J.C. 1993. PCR-based assays of mendelian polymorphisms from anonymous single-copy nuclear DNA: techniques and applications for population genetics. *Mol. Biol. Evol.*, 10, 342-361.

Kearsey, M.J., and Farquhar, A.G.L. 1998. QTL analysis in plants; where are we now? *Heredity*, 80, 137-142.

Kearsey, M.J., and Hyne, V. 1994. QTL analysis: a simple "marker-regression" approach. *Theor. Appl. Genet.*, 89, 698-702.

Keim, P., Diers, B.W., Olson, T.C., and Shoemaker, R.C. 1990. RFLP mapping in soybean: association between marker loci and variation in quantitative traits. *Genetics*, 126, 735-742.

Keim, P., Diers, B.W., and Shoemaker, R.C. 1990. Genetic analysis of soybean hard seededness with molecular markers. *Theor. Appl. Genet.*, 79, 465-469.

Keim, P., Shoemaker, R.C., and Palmer, R.G. 1989. Restriction fragment length polymorphism diversity in soybean. *Theor. Appl. Genet.*, 77, 786-792.

Kennard, W.C., Haveyg, M.J., and Wenzel, G. 1995. Quantitative trait analysis of fruit quality in cucumber: QTL detection, confirmation and comparison with matting design variation. *Theor. Appl. Genet.*, 91, 53-61.

Kennard, W.C., Poetter, K., Dijkhuizen, A., Meglic, V., Staub, J.E., and Havey, M.J. 1994. Linkages among RFLP, RAPD, isozyme, disease-resistance, and morphological markers in narrow and wild crosses of cucumber. *Theor. Appl. Genet.*, 89, 42-48.

Kesseli, R., Witsenboer, H., Stanghellini, M., Vandermark, G., and Michelmore, R. 1993. Recessive resistance to *Plasmopara lactucae-radicis* maps by bulked segregant analysis to a cluster of dominant disease resistance genes in lettuce. *Mol. Plant-Microbe Interact.*, 6, 722-728.

Kesseli, R.V., Paran, I., and Michelmore, R.W. 1994. Analysis of a detailed genetic linkage map of *Lactuca sativa* (lettuce) constructed from RFLP and RAPD markers. *Genetics*, 136, 1435-1446.

Kianian, S.F., and Quiros, C.F. 1992. Generation of a *Brassica oleracea* composite RFLP map: linkage arrangements among various populations and evolutionary implications. *Theor. Appl. Genet.*, 84, 544-554.

Kimura, M. 1980. A simple method for estimating evolutionary rate of base substitutions through comparative studies of nucleotide sequences. *J. Mol. Evol.*, 16, 111-120.

Kisha, T.J., Sneller, C.H., and Diers, B.W. 1997. Relationship between genetic distance among parents and genetic variance in populations of soybean. *Crop Sci.*, 37, 1317-1325.

Kishimoto, N., Higo, H., Abe, K., Arai, S., Siato, A., and Higo. 1994. Identification of the duplicated segments in rice chromosomes 1 and 5 by linkage analysis of cDNA markers of known functions. *Theor. Appl. Genet.*, 88, 722-726.

Kiss, G.B., Csanadi, G., Kalman, K., Kalo, P., and Okresz, L. 1993. Construction of a basic genetic map for alfalfa using RFLP, RAPD, isozyme and morphological markers. *Mol. Gen. Genet.*, 238, 129-137.

Kleinhofs, A., Kilian, A., Saghai Maroof, M.A., Biyashev, R.M., Hayes, P., Chen, F.Q., Lapitan, N., Fenwick, A., Blake, T.K., Kanazin, V., Ananiev, E., Dahleen, L., Kudrna, D., Bollinger, J., Knapp, S.J., Liu, B., Sorrells, M., Heun, M., Franckowiak, J.D., Hoffman, D., Skadsen, R., and Steffenson, B.J. 1993. A molecular, isozyme and morphological map of barley (*Hordeum vulgare*) genome. *Theor. Appl. Genet.*, 86, 705-712.

Knapp, S.J., and Bridges, W.C. 1990. Using molecular markers to estimate quantitative trait locus parameters: power and genetic variances for unreplicated and replicated progeny. *Genetics*, 126, 769-777.

Knapp, S.J., Bridges, W.C., and Birkes, D. 1990. Mapping QTL using molecular marker linkage maps. *Theor. Appl. Genet.*, 79, 583-592.

Koester, R.P., Sisco, P.H., and Stuber, C.W. 1993. Identification of quantitative trait loci controlling days to flowering and plant height in two near isogenic lines of maize. *Crop Sci.*, 33, 1209-1216.

Konieczny, A., and Ausubel, F.M. 1993. A procedure for mapping *Arabidopsis* mutations using codominant ecotype-specific PCR-based markers. *Plant J.*, 4, 403-410.

Korol, A., Ronin, Y., Itzcovich, A., Peng, J., and Nevo, E. 2001. Enhanced efficiency of QTL mapping analysis based on multivariate complexes of quantitative traits. *Genetics*, 157, 1789-1803.

Korol, A., Ronin, Y., and Kirzhner, V. 1995. Interval mapping of quantitative trait loci employing correlated trait complexes. *Genetics*, 140, 1137-1147.

Kosambi, D.D. 1944. The estimation of map distances from recombination values. *Ann. Eugen.*, 12, 172-175.

Kreike, C.M., De Koning, J.R.A., Vinke, J.H., Van Ooijen, J.W., Gebhardt, C., and Stiekema, W.J. 1993. Mapping of loci involved in quantitatively inherited resistance to the potato cyst-nematode *Globodera rostochiensis* pathotype Ro1. *Theor. Appl. Genet.*, 87, 464-470.

Kreike, C.M., Kokwesteneng, A.A., Vinke, J.H., and Stekema, W.J. 1996. Mapping QTLs involved in nematode resistance, tube yield and root development in *Solanum* sp. *Theor. Appl. Genet.*, 92, 463-470.

Kremer, A., Petit, R.J., and Pons, O. 1997. Measures of polymorphism within and between populations. In *Molecular Tools for Screening Biodiversity in Plants and Animals*, Karp, A., Ingram, D.S., and Isaac, P.G., eds., Chapman and Hall, London.

Krug, C.A., Viegas, G.P., and Paolieri, L. 1943. Hibridos comerciais de Milho. *Bragantia*, 3, 367-552.

Kruglyak, L., and Lander, E.S. 1995. A nonparametric approach for mapping quantitative trait loci. *Genetics*, 139, 1421-1428.

Kubisiak, T.L., Nelso, C.D., Nance, W.L., and Stine, M. 1995. RAPD linkage mapping in a longleaf pine × slash pine F_1 family. *Theor. Appl. Genet.*, 90, 1110-1127.

Kurata, N., Moore, G., Nagamura, Y., Foote, T., Yano, M., Minobe, Y., and Gale, M. 1994. Conservation of genome structure between rice and wheat. *BioTechnology*, 12, 276-278.

Lagercrantz, U., Putterill, J., Coupland, G., and Lydiate, D. 1996. Comparative mapping in *Arabidopsis* and *Brassica*, fine scale genome collinearity and congruence of genes controlling flowering time. *Plant J.*, 9, 13-20.

Lagudah, E.S., Appels, R., Brown, A.H.D., and McNeil, D. 1991. The molecular-genetic analysis of *Triticum tauschii*, the D-genome donor to hexaploid wheat. *Genome*, 34, 375-386.

Lamkey, K.R., Hallauer, A.R., and Kahler, A.L. 1987. Allelic differences at enzyme loci and hybrid performance in maize. *J. Hered.*, 78, 231-234.

Lande, R., and Thompson, R. 1990. Efficiency of marker-assisted selection in the improvement of quantitative traits. *Genetics*, 124, 743-756.

Lander, E.S., and Botstein, D. 1989. Mapping mendelian factors underlying quantitative traits using RFLP linkage maps. *Genetics*, 121, 185-199.

Lander, E.S., Green, P., Abrahamson, J., Barlow, A., Daly, M.J., Sincoln, S.E., and Newburg, L. 1987. MAPMAKER: an interactive computer package for constructing primary genetic linkage maps of experimental and natural populations. *Genomics*, 1, 174-181.

Landry, B.S., Hubert, N., Crete, R., Chiang, M.S., Lincoln, S.E., and Etoh, T. 1992. A genetic map for *Brassica oleracea* based on RFLP markers detected with expressed DNA sequences and mapping of resistance genes to race 2 of *Plasmodiophora brassicae* (Woronin). *Genome*, 35, 409-420.

Landry, B.S., Hubert, N., Etoh, T., Harada, J.J., and Lincoln, S.E. 1991. A genetic map for *Brassica napus* based on restriction fragment length polymorphisms detected with expressed DNA sequences. *Genome*, 34, 543-552.

Landry, B.S., Kesseli, R., Farrara, B., and Michelmore, R.W. 1987. A genetic map of lettuce (*Lactuca sativa* L.) with restriction fragment length polymorphism, isozyme, disease resistance and morphological markers. *Genetics*, 116, 331-337.

Lanner-Herrera, C., Gustafsson, M., Falt, A.S., and Bryngelsson, T. 1996. Diversity in natural populations of wild *Brassica oleracea* as estimated by isozyme and RAPD analysis. *Genet. Resources Crop Evol.*, 43, 13-23.

Lark, K.G., Weiseman, J.M., Matthews, B.F., Palmer, R., Chase, K., and Macalma, T. 1993. A genetic map of soybean (*Glycine max* L.) using an intraspecific cross of two cultivars: "Minosy" and "Noir 1". *Theor. Appl. Genet.*, 86, 901-906.

Le Corre, V., Dumolin-Lapègue, and Kremer, A. 1997. Genetic variability at enzyme and RAPD loci in sessile oak *Quercus petraea* (Matt.) Liebl.: the role of history and geography. *Mol. Ecol.*, 6, 519-529.

Lee, M., Godshalk, E.B., Lamkey, K.R., and Woodman, W.W. 1989. Association of restriction fragment length polymorphisms among maize inbred lines with agronomic performance of their crosses. *Crop Sci.*, 29, 1067-1071.

Lee, S.H., Bailey, M.A., Mian, M.A.R., Shipe, E.R., Ashley, D.A., Parrott, W.A., Hussey, R.S., and Boerma, H.R. 1996. Identification of quantitative trait loci for plant height, lodging, and maturity in a soybean population segregating for growth habit. *Theor. Appl. Genet.*, 92, 516-523.

Lefebvre, V., and Palloix, A. 1996. Both epistatic and additive effects of QTLs are involved in polygenic induced resistance to disease: a case study, the interaction pepper-*Phytophthora capsici* Leonian. *Theor. Appl. Genet.*, 93, 503-511.

Lefebvre, V., Palloix, A., Caranta, C., and Pochard, E. 1995. Construction of an intraspecific integrated linkage map of pepper using molecular markers and doubled haploid progenies. *Genome*, 38, 112-121.

Lefort-Buson, M. 1985a. Distance genetique et heterosis: 1. Mise en evidence d'une relation entre heterosis et divergence genetique. In *Les Distances Genetiques*. Lefort-Buson, M., and de Vienne, D., eds., INRA, Paris.

Lefort-Buson, M. 1985b. Distance genetique et heterosis: 4. Utilisation des criteres biometriques. In *Les Distances Genetiques*. Lefort-Buson, M., and de Vienne, D., eds., INRA, Paris.

Leonards-Schippers, C., Gieffers, W., Schafer-Pregl, R., Ritter, E., Knapp, S.J., Salamini, F., and Gebhardt, C. 1994. Quantitative resistance to *Phytophthora infestans* in potato: a case study for QTL mapping in allogamous species. *Genetics*, 137, 67-77.

Lerman, L.S., and Silverstein, K. 1987. Computational simulation of DNA melting and its application to denaturing gradient gel electrophoresis. *Methods Enzymol.*, 155, 482-501.

Lewontin, R.C. 1972. The apportionment of human diversity. *Evol. Biol.*, 6, 381-398.

Leyser, H.M.O., Lincoln, C.A., Timpte, C., Lammer, D., Turner, J., and Estelle, M. 1993. *Arabidopsis* auxin-resistance gene AXR1 encodes a protein related to ubiquitin-activating enzyme E1. *Nature*, 364, 161-164.

Li, Z., Pinson, S.R.M., Stansel, J.W., and Park, W.D. 1995. Identification of quantitative trait loci (QTLs) for heading date and plant height in cultivated rice (*Oryza sativa* L.). *Theor. Appl. Genet.*, 91, 374-379.

Lin, H.X., Qian, H.R., Zhuang, J.Y., Lu, J., Min, S.K., Xiong, Z.M., Huang, N., and Zheng, K.L. 1996. RFLP mapping of QTLs for yield and related characters in rice (*Oryza sativa* L.). *Theor. Appl. Genet.*, 92, 920-927.

Lin, Y.R., Schertz, K.F., and Paterson, A.H. 1995. Comparative analysis of QTLs affecting plant height and maturity across the Poaceae, in reference to an interspecific sorghum population. *Genetics*, 141, 391-411.

Lisitsyn, N.A., Lisitsina, N., and Wigler, M. 1993. Cloning the differences between two complex genomes. *Science*, 259, 946-951.

Liu, B.H., and Knapp, S.J. 1990. GMENDEL: a program for mendelian segregation and linkage analysis of individual or multiple progeny populations using log-likelihood ratios. *J. Hered.*, 81, 407.

Liu, C.J., Witcombe, J.R., Pittaway, T.S., Nash, M., Hash, C.T., Busso, C.S., and Gale, M.D. 1994. An RFLP-based genetic map of pearl millet (*Pennisetum glaucum*). *Theor. Appl. Genet.*, 89, 481-487.

Liu, S.C., Kowalski, S.P., Lan, T.H., Feldmann, K.A., and Paterson, A.H. 1996. Genome-wide high-resolution mapping by recurrent intermating using *Arabidopsis thaliana* as a model. *Genetics*, 142, 247-258.

Liu, Z, and Furnier, G.R. 1993. Comparison of allozyme, RFLP and RAPD markers for revealing genetic variation within and between trembling aspen and bigtooth aspen. *Theor. Appl. Genet.*, 87, 97-105.

Livingstone, K.D., Lackney, V.K., Blauth, J.R., van Wijk, R., and Jahn M.K. 1999. Genome mapping in *Capsicum* and the evolution of genome structure in the Solanaceae. *Genetics*, 152, 1183-1202.

Livini, C., Ajmone-Marsan, P., Melchinger, A.E., Messmer, M.M., and Motto, M. 1992. Genetic diversity of maize inbred lines within and among heterotic groups revealed by RFLPs. *Theor. Appl. Genet.*, 84, 17-25.

Lukens L., Doebley J. 1999. Epistatic and environmental interactions for quantitative trait loci involved in maize evolution. *Genetical Research*, 1999, 74, 291-302

Luo, Z.W., Thompson, R., and Woolliams, J.A. 1997. A population genetics model of marker-assisted selection. *Genetics*, 146, 1173-1183.

Luro, F., Lorieux, M., Laigret, F., Bove, J.M., and Ollitrault, P. 1995. Cartographie du genome des agrumes a l'aide des marqueurs moleculaires et distorsions de segregation. In *Techniques et Utilisations des Marqueurs Moleculaires*, Berville, A., and Tersac, M., eds., Montpellier, 29-31 Mar. 1994, INRA, Paris, "Les Colloques", 72, 69-82.

Lynch, M. 1990. The similarity index and DNA fingerprinting. *Mol. Biol. Evol.*, 7, 478-484.

Lynch, M., and Crease, T.J. 1990. The analysis of population survey data on DNA sequence variation. *Mol. Biol. Evol.*, 7, 377-394.

Lynch, M., and Milligan, B.G. 1994. Analysis of population genetic structure with RAPD markers. *Mol. Ecol.*, 3, 91-99.

Lyon, M.F. 1990. Dunn and mouse genetic mapping. *Genetics*, 125, 231-236.

Maisonneuve, B., Bellec, Y., Anderson, P., and Michelmore, R.W. 1994. Rapid mapping of two genes for resistance to downy mildew from *Lactuca serriola* to existing cluster of resistance genes. *Theor. Appl. Genet.*, 89, 96-104.

Mamuris, Z., Stamatis, C., and Triantaphyllidis, C. 1999. Intraspecific genetic variation of striped red mullet (*Mullus surmuletus* L.) in the Mediterranean Sea assessed by allozyme and random amplified polymorphic DNA (RAPD) analysis. *Heredity*, 83, 30-38.

Mangin, B., Goffinet, B., and Rebai, A. 1994. Constructing confidence intervals for QTL location. *Genetics*, 138, 301-1308.

Manjarrez-Sandoval, P., Carter, T.E.J., Webb, D.M., and Burton, J.W. 1997. RFLP genetic similarity estimates and coefficient of parentage as genetic variance predictors for soybean yield. *Crop Sci.*, 37, 698-703.

Mansur, L.M., Lark, K.G., Kross, H., and Oliviera, A. 1993a. Interval mapping of quantitative trait loci for reproductive, morphological and seed traits of soybean (*Glycine max* L.). *Theor. Appl. Genet.*, 86, 907-913.

Mansur, L.M., Orf, J., and Lark, K.G. 1993b. Determining the linkage of quantitative trait loci to RFLP markers using extreme phenotypes of recombinant inbreds of soybean (*Glycine max* L. Merr.). *Theor. Appl. Genet.*, 86, 914-918.

Mariette S., Chagné D., Lézier C., Pastuszka P., Raffin A., Plomion C., and Kremer A. 2001. Genetic diversity within and among *Pinus pinaster* populations: comparison between AFLP and microsatellite markers. *Heredity*, 86, 469-479.

Mariette, S., Le Corre, V., Austerlitz, F., and Kremer, A. 2002a. Sampling within the genome for measuring within-population diversity: trade-offs between markers. *Mol. Ecol.* (submitted).

Mariette, S., Cottrell, J., Csaikl, U., Goikoechea, P., König, A., Lowe, A.J., Van dam, B.C., Barrenche, T., Bodénès, C., Streiff, R., Burg, K., Groppe, K., Munro, R.C., Tabbener, H., and Kremer, A. 2002b. Comparison of levels of genetic diversity detected with AFLP and microsatellite marker within and among mixed *Q. petraea* (Matt.) Liebl. and *Q. robur* L. stands. *Silvae Genetica* (in press).

Marshall, D.R., and Jain, S.K. 1969. Genetic polymorphism in natural populations of *Avena fatua* and *A. barbata. Nature*, 221, 276-278.

Marshall, T.C., Slate, J., Kruuk, L.E.B., and Pemberton, J.M. 1998. Statistical confidence for likelihood-based paternity inference in natural populations. *Mol. Ecol.*, 7, 639-655.

Martin, G.B., Brommonschenkel, S.H., Chunwongse, J., Frary, A., Ganal, M.W., Spivey, R., Wu, T., Earle, E.D., and Tanksley, S.D. 1993. Map-based cloning of a protein kinase gene conferring disease resistance in tomato. *Science*, 262, 1432-1436.

Martin, G.B., Williams, J.G.K., and Tansksley, S.D. 1991. Rapid identification of markers linked to a *Pseudomonas* resistance gene in tomato by using random primers and near-isogenic lines. *Proc. Natl. Acad. Sci. USA*, 88, 2336-2340.

Martin, J.M., Talbert, L.E., Lanning, S.P., and Blake, N.K. 1995. Hybrid performance in wheat as related to parental diversity. *Crop Sci.*, 35, 104-108.

Martinez, O., and Curnow, R.N. 1992. Estimating the locations and the sizes of the effects of quantitative trait loci using flanking markers. *Theor. Appl. Genet.*, 85, 480-488.

Mason-Gamer, R.J., Holsinger, K.E., and Jansen, R.K. 1995. Chloroplast DNA haplotype variation within and among populations of *Coreopsis grandi-flora* (Asteraceae). *Mol. Biol. Evol.*, 12, 371-381.

Maughan, P.J., Maroof, M.A.S., and Buss, G.R. 1996a. Molecular-marker analysis of seed-weight: genomic locations, gene action, and evidence for orthologous evolution among three legume species. *Theor. Appl. Genet.*, 93, 574-579.

Maughan, P.J., Saghai Maroof, M.A., Buss, G.R., and Huestis, G.M. 1996b. Amplified fragment length polymorphism (AFLP) in soybean: species diversity, inheritance, and near-isogenic line analysis. *Theor. Appl. Genet.*, 93, 392-401.

McCauley, D.E. 1995. The use of chloroplast DNA polymorphism in studies of gene flow in plants. *Trends Ecol. Evol.*, 10, 198-202.

McCauley, D.E., Raveill, J., and Antonovics, J. 1995. Local founding events as determinants of genetic structure in a plant metapopulation. *Heredity*, 75, 630-636.

McCouch, S.R., Kochert, G., Yu, Z.H., Wang, Z.Y., Kush, G.S., Coffman, W.R., and Tanksley, S.D. 1988. Molecular mapping of rice chromosomes. *Theor. Appl. Genet.*, 76, 815-829.

McDonald, J.H. 1994. Detecting natural selection by comparing geographic variation in protein and DNA polymorphisms. In *Non-neutral Evolution. Theories and Molecualr Data*, Golding, B., ed., Chapman and Hall, pp. 88-101.

McGrath, J.M., and Quiros, C.F. 1992. Genetic diversity at isozyme and RFLP loci in *Brassica campestris* as related to crop type and geographical origin. *Theor. Appl. Genet.*, 83, 783-790.

Meagher, T.R., and Thompson, E. 1986. The relationship between single parent and parent pair genetic likelihoods in genealogy reconstruction. *Theor. Pop. Biol.* 29, 87-106.

Meksem, K., Leister, D., Peleman, J., Zabeau, M., Salamini, F., and Gebhardt, C. 1995. A high resolution map of the vicinity of the R1 locus on chromosome V of potato based on RFLP and AFLP markers. *Mol. Gen. Genet.*, 249, 74-81.

Meksem, K., Ruben, E., Hyten, D., Triwitayakorn, K., and Lightfoot, D.A. 2001. Conversion of AFLP bands into high-throughput DNA markers. *Mol. Genet. Genomics*, 265, 207-214.

Melake Berhan, A., Hulbert, S.H., Butlmer, L.G., and Bennetzen, J.L. 1993. Structure and evolution of the genomes of sorghum bicolor and *Zea mays*. *Theor. Appl. Genet.*, 86, 598-604.

Melchinger, A.E. 1999. Genetic diversity and heterosis. In *"The Genetics and Exploitation of Heterosis in Crops."* J.G. Coors and S. Pandey Ed. International Symposium on the Genetics and Exploitation of Heterosis in Crops, CIMMYT, Mexico, 17-22/08/1997, chap. 10, 99-118.

Melchinger, A.E., Boppenmeier, J., Dhillon, B.S., Pollmer, W.G., and Herrmann, R.G. 1992. Genetic diversity for RFLPs in European maize inbreds: II. Relation to performance of hybrids within versus between heterotic groups for forage traits. *Theor. Appl. Genet.*, 84, 672-681.

Melchinger, A.E., Graner, A., Singh, M., and Messmer, M.M. 1994. Relationships among European barley germplasm: I. Genetic diversity among winter and spring cultivars revealed by RFLPs. *Crop Sci.*, 34, 1191-1199.

Melchinger, A.E., Lee, M., Lamkey, K.R., Hallauer, A.R., and Woodman, W.L. 1990a. Genetic diversity for restriction fragment length polymorphisms and heterosis for two diallel sets of maize inbreds. *Theor. Appl. Genet.*, 80, 488-496.

Melchinger, A.E., Lee, M., Lamkey, K.R., and Woodman, W.L. 1990b. Genetic diversity for restriction fragment length polymorphisms: relation to estimated genetic effects in maize inbreds. *Crop Sci.*, 30, 1033-1040.

Melchinger, A.E., Messmer, M.M., Lee, M., Woodman, W.L., and Lamkey, K.R. 1991. Diversity and relationships among U.S. maize inbreds revealed by restriction fragment length polymorphisms. *Crop Sci.*, 31, 669-678.

Menancio-Hautea, D., Fatokun, C.A., Kumar, L., Danesh, D., and Young, N.D. 1993. Comparative genome analysis of mungbean (*Vigna radiata* L. Wilczek) and cowpea (*V. unguiculata* L. Walpers) using RFLP mapping data. *Theor. Appl. Genet.*, 86, 797-810.

Messmer, M.M., Melchinger, A.E., Boppenmeier, J., Herrmann, R.G., Brunkslaus-Jung, E. 1992. RFLP analyses of early-maturing European maize germplasm. I. Genetic diversity among flint and dent inbreds. *Theor. Appl. Genet.*, 83, 1003-1012.

Michalakis, Y., and Excoffier, L. 1996. A generic estimation of population subdivision using distances between alleles with special reference for microsatellite loci. *Genetics*, 142, 1061-1064.

Michelmore, R. 1995a. Molecular approaches to manipulation of disease resistance genes. *Ann. Rev. Phytopathol.*, 15, 393-427.

Michelmore, R. 1995b. Isolation of disease resistance genes from crop plants. *Curr. Opin. Biotech.*, 6, 145-152.

Michelmore, R.W., Paran, I., and Kesseli, R.V. 1991. Identification of markers linked to disease resistance genes by bulked segregant analysis: a rapid method to detect markers in specific genomic regions using segregating populations. *Proc. Natl. Acad. Sci. USA*, 88, 9828-9832.

Moll, R.H., Lonnquist, J.H., Fortuna, J.V., and Johnson, E.C. 1965. The relationship of heterosis and genetic divergence in maize. *Genetics*, 52, 139-144.

Moore, G., Devos, K.M., Wang, Z., and Gale, M.D. 1995. Cereal genome evolution. Grasses, line up and form a circle. *Curr. Biol.*, 5, 737-739.

Moran, G.F. 1992. Patterns of genetic diversity in Australian tree species. *New Forests*, 6, 49-66.

Moreau, L., Charcosset, A., Hospital, F., and Gallais, A. 1998. Marker-assisted selection efficiency in populations of finite size. *Genetics*, 148, 1353-1355.

Moreau, L., Lemarié, S., Charcosset, A., and Gallais, A. 2000. Economic efficiency of marker assisted selection. *Crop Sci.*, 40, 329-337.

Morgan, T.H. 1911. Random segregation versus coupling in mendelian inheritance. *Science,* 36, 718-719.

Morgante, M., and Olivieri, A.M. 1993. PCR-amplified microsatellites as markers in plant genetics. *Plant J.*, 3, 175-182.

Morgante, M., Rafalski, J.A., Biddle, P., Tingey, S., and Olivieri, A.M. 1994. Genetic mapping and variability of seven soybean simple sequence repeat loci. *Genome*, 37, 763-769.

Morizot, D.C. 1990. Use of fish gene maps to predict ancestral vertebrate genome organization. In *Isozymes: Structure, Function and Use in Biology and Medicine*, Ogita, Z.I., and Marker, C.L., eds., Liss, Wiley, New York, pp. 207-234.

Morton, N.E. 1955. Sequential tests for the detection of linkage. *Am. J. Hum. Genet.*, 7, 277-318.

Moser, H., and Lee, M. 1994. RFLP variation and genealogical distance, multivariate distance, heterosis, and genetic variance in oats. *Theor. Appl. Genet.*, 87, 947-956.

Murigneux, A., Barloy, D., Leroy, P., and Beckert, M. 1993. Molecular and morphological evaluation of doubled-haploid lines in maize. 1. Homogeneity within DH lines. *Theor. Appl. Genet.*, 86, 837-842.

Mutschler, M.A., Doerge, R.W., Liu, S.C., Kuai, J.P., Liedl, B.E., and Shapiro, J.A. 1996. QTL analysis of pest resistance in the wild tomato *Lycopersicon pennellii*: QTLs controlling acylsugar level and composition. *Theor. Appl. Genet.*, 92, 709-718.

Myers, R.M., Maniatis, T., and Lerman, S. 1987. Detection and localization of single base changes by denaturing gradient gel electrophoresis. *Methods Enzymol.*, 155, 501-527.

Nadler, S.A., Lindquist, R.L., and Near, T.J. 1995. Genetic structure of midwestern *Ascaris suum* populations: a comparison of isoenzyme and RAPD markers. *J. Parasitol.*, 81, 385-394.

Nam, H.G., Giraudat, J., den Boer, B., Moonan, F., Loos, W.D.B., Hauge, B.M., and Goodman, H.M. 1989. Restriction fragment length polymorphism linkage map of *Arabidopsis thaliana*. *Plant Cell*, 1, 699-705.

Neale, D.B., Saghai Maroof, M.A., Allard, R.W., Zhang, Q., and Jorgensen, R.A. 1988. Chloroplast DNA diversity in populations of wild and cultivated barley. *Genetics*, 120, 1105-1110.

Nei, M. 1973. Analysis of gene diversity in subdivided populations. *Proc. Natl. Acad. Sci. USA*, 70, 3321-3323.

Nei, M., 1982. Evolution of human races at the gene level. In "Human genetics. Part A. The unfolding genome", B. Bonne-Tamir, T. Cohen and P.M. Goodman eds, Alan R. Liss, New York, pp 167-181.

Nei, M. 1987. *Molecular Evolutionary Genetics*. Columbia University Press.

Nei, M., and Chesser, R.K. 1983. Estimation of fixation indices and gene diversities. *Ann. Hum. Genet.*, 47, 253-259.

Nei, M., and Li, W.H. 1979. Mathematical model for studying genetic variation in terms of restriction endonucleases. *Proc. Natl. Acad. Sci. USA*, 76, 5269-5273.

Nei, M., and Miller, J.C. 1990. A simple method for estimating average number of nucleotide substitutions within and between populations from restriction data. *Genetics*, 125, 873-879.

Nei, M., and Roychoudhury, A.K. 1974. Sampling variances of heterozygosity and genetic distance. *Genetics*, 76, 379-380.

Nelson, C.D., Nance, W.L., and Doudrick, R.L. 1993. A partial genetic linkage map of slash pine (*Pinus elliotti* Englem var. *elliottii*) based on random amplified polymorphic DNAs. *Theor. Appl. Genet.*, 87, 145-151.

Nevo, E., Beiles, A., and Ben-Shlomo, R. 1984. The evolutionary significance of genetic diversity: ecological, demographic and life history correlates. In *Evolutionary Dynamics of Genetic Diversity*, Mani, G.S., ed., Springer-Verlag (Lecture Notes in Biomathematics, no. 53), pp. 13-213.

Nienhuis, J., and Sills, G. 1992. The potential of hybrid varieties in self-pollinating vegetables. In *Reproductive Biology and Plant Breeding*, Proc. XIIIth Eucarpia Meeting. Dattee, Y., Dumas, C., and Gallais, A., eds., Springer-Verlag.

Nodari, R.O., Tsai, S.M., Gilbertson, R.L., and Gepts, P. 1993. Towards an integrated linkage map of common bean. II. Development of an RFLP-based linkage map. *Theor. Appl. Genet.*, 85, 513-520.

Nybom, H., and Bartish, I.V. 2000. Effects of life history traits and sampling strategies on genetic diversity estimates obtained with RAPD markers in plants. *Perspectives in Plant Ecol., Evol. Systematics*, 3/2, 93-114.

O'Donoughue, L.S., Wang, Z., Roder, M., Keen, B., Legget, M., Sorrells, M.E., and Tanksley, S.D. 1992. An RFLP-based linkage map of oats based on cross between two diploid taxa (*Avena atlantica* × *A. hirtula*). *Genome*, 35, 765-771.

Ohta, T., and Kimura, M. 1973. A model of mutation appropriate to estimate the number of electrophoretically detectable alleles in a finite population. *Genet. Res. Camb.*, 22, 201-204.

Oliver S.G., van der Aart Q.J., Agostoni-Carbone M.L., Aigle M., Alberghina L., Alexandraki D., Antoine G., Anwar R., Ballesta J.P., Benit P., et al. 1992. The complete DNA sequence of yeast chromosome III. *Nature*, 357, 38-46.

Olson, M., Hood, L., Cantor, C., and Doststein, D. 1989. A common language for physical mapping of the human genome. *Science*, 254, 1434-1435.

Orita, M., Iwahana, H., Kanasawa, H., Hayashi, K., and Sekiya, T. 1989. Detection of polymorphisms of human DNA by gel electrophoresis as single-strand conformation polymorphism. *Proc. Natl. Acad. Sci. USA*, 86, 2766-2770.

Paglia, G.P., Olivieri, A.M., and Morgante, M. 1998. Towards second-generation STS (sequence-tagged sites) linkage maps in conifers: a genetic map of norway spruce (*Picea abies* K.). *Mol. Gen. Genet.*, 258, 466-478.

Pammi, S., Schertz, K., Xu, G., Hart, G., and Mullet, J.E. 1994. Random-amplified-polymorphic DNA markers in sorghum. *Theor. Appl. Genet.*, 89, 80-88.

Pan, A., Hayes, P.M., Chen, T.H.H., Blake, T., Wright, S., Karsai, I., and Bedo, Z. 1994. Genetic analysis of the components of winter hardiness in barley (*Hordeum vulgare* L.). *Theor. Appl. Genet.*, 89, 900-910.

Panaud, O., Chen, X., and McCouch, S.R. 1995. Frequency of microsatellite sequences in rice (*Oryza sativa* L.). *Genome*, 38, 1170-1176.

Papa, R., Attene, G., Barcaccia, G., Ohgata, A., and Konishi, T. 1998. Genetic diversity in landrace populations of *Hordeum vulgare* L. from Sardinia, Italy, as revealed by RAPDs, isozymes and morphophenological traits. *Plant Breeding*, 117, 523-530.

Paran, I., Goldman, I.L., Tanksley, S.D., and Zamir, D. 1995. Recombinant inbred lines for genetic mapping in tomato. *Theor. Appl. Genet.*, 90, 542-548.

Paran, I., and Michelmore, R.W. 1993. Development of reliable PCR-based markers linked to downy mildew resistant genes in lettuce. *Theor. Appl. Genet.*, 85, 985-993.

Paran, I., Kesseli, R., and Michelmore, R. 1991. Identification of restriction fragment length polymorphism and random amplified polymorphic DNA markers linked to downy mildew resistance genes in lettuce, using near-isogenic lines. *Genome*, 34, 1021-1027.

Paterson, A.H., Damon, S., Hewitt, J.D., Zamir, D., Rabinowitch, H.D., Lincoln, S.E., Lander, E.S., and Tanksley, S.D. 1991. Mendelian factors underlying quantitative traits in tomato: comparison across species, generations and environments. *Genetics*, 127, 181-187.

Paterson, A.H., De Verna, J.W., Lanini, B., and Tanksley, S.D. 1990. Fine mapping of quantitative trait loci using selected overlapping recombinant chromosomes, in an interspecific cross of tomato. *Genetics*, 124, 735-742.

Paterson, A.H., Lander, E.S., Hewitt, J.D., Peterson, S., Loncoln, S.E., and Tanksley, S.D. 1988. Resolution of quantitative traits into Mendelian factors by using a complete linkage map of restriction fragment length polymorphisms. *Nature*, 335, 721-726.

Paterson, A.H., Lin, Y.R., Li, Z, Schertz, K.F., Doebley, J.F., Pinson, S.R., Liu, S.C., Stansel, J.W., and Irvine, J.E. 1995. Convergent domestication of cereal crops by independent mutations at corresponding genetic loci. *Science*, 269, 1714-1718.

Pe, M.E., Gianfranceschi, L., Taramino, G., Tarchini, G., Angelini, P., Dani, M., and Binelli, G. 1993. Mapping quantitative trait loci (QTLs) for resistance to *Gibberella zeae* infection in maize. *Mol. Gen. Genet.*, 241, 11-16.

Peakall, P., Smouse, P.E., and Huff, D.R. 1995. Evolutionary implications of allozyme and RAPD variation in diploid populations of dioecious buffalograss *Buchloe dactyloides. Mol. Ecol.*, 4, 135-147.

Peltier, D., Farcy, E., Dulieu, H., Berville, A. 1994. Origin, distribution and mapping of RAPD markers from wild *Petunia* species in *Petunia hybrida* Hor lines. *Theor. Appl. Genet.*, 88, 637-645.

Peng, J.Y., Glaszmann, J.C., and Virmani, S.S. 1988. Heterosis and isozyme divergence in indica rice. *Crop Sci.*, 28, 561-563.

Penner, G.A., Chong, J., Levesque-Lemay, Molnar, S.J., and Fedak, G. 1993. Identification of a RAPD marker linked to the oat stem rust gene *PG3. Theor. Appl. Genet.*, 85, 702-705.

Pereira, M.G., and Lee, M. 1995. Identification of genomic regions affecting plant height in sorghum and maize. *Theor. Appl. Genet.*, 90, 380-388.

Petit, R.J., Kremer, A., and Wagner, D.B. 1993a. Finite island model for organelle and nuclear genes in plants. *Heredity*, 71, 630-641.

Petit, R.J., Kremer, A., and Wagner, D.B. 1993b. Geographic structure of chloroplast DNA polymorphism in European oaks. *Theor. Appl. Genet.*, 87, 122-128.

Petit, R.J., Brewer, S., Bordács, S., Burg, K., Cheddadi, R., Coart, E., Cottrell, J., Csaikl, U.M., van Dam, B.C., Deans, J.D., Fineschi, S., Finkeldey, R., Glaz, I., Goicoechea, P.G., Jensen, J.S., König, A.O., Lowe, A.J., Madsen, S.F., Mátyás, G., Munro, R.C., Popescu, F., Slade, D., Tabbener, H., de Vries, S.M.G., Ziegenhagen, B., de Beaulieu, J.L., and Kremer, A. 2002. Identification of refugia and postglacial colonization routes of European white oaks based on chloroplast DNA and fossil pollen evidence. *Forest Ecol. Mgmt.* (in press).

Pflieger, S., Lefebvre, V., and Causse, M. 2001. The candidate gene approach in plant genetics: a review. *Mol. Breeding*, 7, 275-291.

Philipp, U., Wehling, P., and Wricke, G. 1994. A linkage map of rye. *Theor. Appl. Genet.*, 88, 243-248.

Pielou, E.C. 1975. *Ecological Diversity*. Wiley, New York.

Plaschke, J., Boerner, A., Xie, D.X., Koebner, R.M.D., Schlegel, R., and Gale, M.D. 1993. RFLP mapping of genes affecting plant height and growth habit in rye. *Theor. Appl. Genet.*, 85, 1049-1054.

Plomion, C., Bahrman, D., Durel, C.E., and O'Malley, D.M. 1995a. Genomic mapping in *Pinus pinaster* (maritime pine) using RAPD and protein markers. *Heredity*, 74, 661-668.

Plomion, C., O'Malley, D.M., and Durel, C.E. 1995b. Genomic analysis in maritime pine (*Pinus pinaster*). Comparison of two rapid maps using selfed and open-pollinated seeds of the same individual. *Theor. Appl. Genet.*, 90, 1028-1034.

Pogson, G.H., Mesa, K.A., and Boutilier, R.G. 1995. Genetic population structure and gene flow in the Atlantic cod *Gadus morhua*: a comparison of allozyme and nuclear RFLP loci. *Genetics*, 139, 375-385.

Pons, O., and Chaouche, K. 1996. Estimation, variance and optimal sampling strategy of gene diversity. II. Diploid locus. *Theor. Appl. Genet.*, 91, 122-130.

Pons, O., and Petit, R.J. 1995. Estimation, variance and optimal sampling strategy. 1. Haploid locus. *Theor. Appl. Genet.*, 90, 462-470.

Pons, O., and Petit, R.J. 1996. Measuring and testing genetic differentiation with ordered versus unordered alleles. *Genetics*, 144, 1237-1245.

Powell, J.R. 1994. Molecular techniques in population genetics: a brief history. In *Molecular Ecology and Evolution: Approaches and Applications*, Schierwater, B., Streit, B., Wagner, G.P., and Desalle, R., eds., Birkhauser Verlag, pp. 131-157.

Powell, W., Morgante, M., Andre, C., Hanafey, M., Vogel, J., Tingey, S., and Rafalski, J.A. 1996. The comparison of RFLP, RAPD, AFLP and SSR (microsatellite) markers for germplasm analysis. *Mol. Breeding*, 2, 225-238.

Powell, W., Morgante, M., McDewitt, R., Vendramin, G.G., and Rafalski, J.A. 1995. Polymorphic simple sequence repeat regions in chloroplast genomes: applications to the population genetics of pines. *Proc. Natl. Acad. Sci. USA*, 92, 7759-7763.

Price, S.C., Kahler, A.L., Hallauer, A.R., Charmley, P., and Giegel, D.A. 1986. Relationship between performance and multilocus heterozygosity at enzyme loci in single-cross hybrids of maize. *J. Hered.*, 77, 341-344.

Prince, J.P., Pochard, E., and Tanksley, S.D. 1993. Construction of a molecular linkage map of pepper and a comparison of synteny with tomato. *Genome*, 36, 404-417.

Prioul, J.L., Quarrie, S., Causse, M., and de Vienne, D. 1997. Dissecting complex physiological functions into elementary components through the use of molecular quantitative genetics. *J. Exp. Bot.*, 48, 1151-1163.

Ragab, R.A., Dronavalli, S., Saghai Maroof, M.A., and Yu, Y.G. 1994. Construction of a sorghum RFLP linkage map using sorghum and maize DNA probes. *Genome*, 37, 590-594.

Ragot M., Biasiolli M., Delbut M.F., Dell'orco A., Malgarini L. Thevenin P., Vernoy J., Vivant J., Zimmermann R., Gay G., 1995. Marker-assisted backcrossing : a pratical example. In *Techniques et utilisations des marqueurs moléculaires*. Ed. INRA Paris, p. 45-46.

Rajapakse, S., Belthoff, L.E., He, G., Estager, A.E., Scorza, R., Verde, I., Ballard, R.E., Baird, W.V., Callahan, A., Monet, R., and Abbott, A.G. 1995. Genetic linkage mapping in peach using morphological RFLP and RAPD markers. *Theor. Appl. Genet.*, 90, 503-510.

Ray, J.D., Yu, L., McCouch, S.R., Champoux, M.C., Wang, G., and Nguyen, H.T. 1996. Mapping quantitative trait loci associated with root penetration ability in rice (*Oryza sativa* L). *Theor. Appl. Genet.*, 92, 627-636.

Raybould, A.F., Mogg, R.J., Clarke, R.T., Gliddon, C.J., and Gray, A.J. 1999. Variation and population structure at microsatellite and isozyme loci in wild cabbage (*Brassica oleracea* L.) in Dorset (UK). *Genet. Resources Crop Evol.*, 46, 351-360.

Rebai, A. 1997. Comparison of methods for regression interval mapping in QTL analysis with non-normal traits. *Genet. Res. Camb.*, 69, 69-74.

Rebai, A., Goffinet, B., and Mangin, B. 1994. Approximate thresholds of interval mapping tests for QTL detection. *Genetics*, 138, 235-240.

Rebai, A., Goffinet, B., and Mangin, B. 1995. Comparing power of different methods for QTL detection. *Biometrics*, 51, 87-99.

Reboud, X., and Zeyl, C. 1994. Organelle inheritance in plants. *Heredity*, 72, 132-140.

Redona, E.D., and Mackill, D.J. 1996. Mapping quantitative trait loci for seedling vigor in rice using RFLPs. *Theor. Appl. Genet.*, 92, 395-402.

Reiter, R.S., Coors, J.G., Sussman, M.R., and Gabelman, W.H. 1991. Genetic analysis of tolerance to low phosphorus stress in maize using RFLPs. *Theor. Appl. Genet.*, 82, 561-568.

Reiter, R.S., Williams, J., Feldman, K., Rafalski, J.A., Tingey, S.V., and Scolnik, P.A. 1992. Global and local genome mapping in *Arabidopsis thaliana* recombinant inbred lines and random amplified polymorphic DNAs. *Proc. Natl. Acad. Sci. USA*, 89, 1477-1481.

Remington D.L., Whetten R.W., Liu B.H., and O'Malley D.M. 1999. Construction of an AFLP genetic map with nearly complete genome coverage in *Pinus taeda*. *Theor. Appl. Genet.*, 98, 1279-1292.

Ribaut, J.M., Hoisington, D.A., Deutsch, J.A., Jiang, C., Gonzalez de Leon, D. 1996. Identificaiton of quantitative trait loci under drougth conditions in tropical maize. 1. Flowering parameters and the anthesis-silking interval. *Theor. Appl. Genet.*, 92, 905-914.

Rick, C.M. 1969. Controlled introgression of chromosomes of *Solanum pennellii* into *Lycopersicon esculentum*: segregation and recombination. *Genetics*, 62, 753-768.

Ritter, E., Gebhardt, C., and Salamini, F. 1990. Estimation of recombination frequencies and construction of RFLP linkage maps in plant from crosses between heterozygous parents. *Genetics*, 125, 645-654.

Robert, L.S., Robson, F., Sharpe, A., Lydiate, D., and Coupland, G. 1998. Conserved structure and function of the *Arabidopsis* flowering time gene CONSTANS in *Brassica napus*. *Plant Mol. Biol.*, 37, 763-772.

Robertson, D.S. 1984. Different frequency in the recovery of crossover products from male and female gametes of plants hypoploid for B-A translocation in maize. *Genetics*, 107, 117-130.

Rodolphe, F., and Lefort, M. 1993. A multi-marker model for detecting chromosomal segments displaying QTL activity. *Genetics*, 134, 1277-1288.

Rogers, J.S. 1972. Measures of genetic similarity and genetic distance. In *Studies in Genetics*, VII, Wheeler, M.R., ed., University of Texas Publ. 7213, pp. 145-153.

Romagosa, I., Ullrich, S.E., Han, F., and Hayes, P.M. 1996. Use of the additive main effects and multiplicative interaction model in QTL mapping for adaptation in barley. *Theor. Appl. Genet.*, 93, 30-37.

Rommens, J.M., Iannuzzi, M.C., Kerem, B.S., Drumm, M.L., Melmer, G., Dean, M., Rozmahel, R., Cole, J.L., Kennedy, D., Hidaka, N., Zsiga, M., Buchwald, M., Riordan, J.R., Tsui, L.C., and Collins, F.S. 1989. Identification of the cystic fibrosis gene: chromosome walking and jumping. *Science*, 245, 1059-1065.

Rongwen, J., Akkaya, M.S., Bhagwat, A.A., Lavi, U., and Cregan, P.B. 1995. The use of microsatellite DNA markers for soybean genotype identification. *Theor. Appl. Genet.*, 90, 43-48.

Ross, K.G., Krieger, M.J.B., Dewayne Shoemaker, D., Vargo, E.L., and Keller, L. 1997. Hierarchical analysis of genetic structure in native fire ant populations: results from three classes of molecular markers. *Genetics*, 147, 643-655.

Rowland, L.J., and Levi, A. 1994. RAPD-based genetic linkage map of blueberry derived from a cross between diploid species (*Vaccinium darrowi* and *V. elliotti*). *Theor. Appl. Genet.*, 87, 863-868.

Saghai-Maroof, M.A., Biyashev, R.M., Yang, G.P., Zhang, Q., and Allard, R.W. 1994. Extraordinarily polymorphic microsatellite DNA in barley: species diversity, chromosomal locations and population dynamics. *Proc. natl. Acad. Sci. USA*, 91, 5466-5470.

Saghai-Maroof, M.A., Yue, Y.G., Xiang, Z.X., Stromberg, E.L., and Rufener, G.K. 1996. Identification of quantitative trait loci controlling resistance to gray leaf spot disease in maize. *Theor. Appl. Genet.*, 93, 539-546.

Saghai Maroof, M.A., G.P. Yang, et al. 1997. "Correlation between molecular marker distance and hybrid performance in U.S. southern long grain rice." *Crop Science* 37: 145-150.

Saiki, R.K., Gelfland, D.H., Stoffel, S., Scharf, S.J., Higuchi, R., Horn, G.T., Mullis, K.B., and Ehrlich, H.A. 1988. Primer-directed enzymatic amplification of DNA with thermostable DNA polymerase. *Science*, 239, 487-491.

Saito, A., Yano, M., Kishimoto, N., Nakagahra, M., Yoshimura, A., Saito, K., Kuhara, S., Ukai, Y., Kawase M., Nagamine, T., Yoshimura, S, Ideta, O., Ohsawa, R., Hayano, Y. Iwata, N., and Sigiura, M. 1991. Linkage map of restriction fragment length polymorphism loci in rice. *Jpn. J. Breed.*, 41, 665-670.

Sarfatti, M., Katan, J., Fluhr, R., and Zamir, D. 1989. An eflp marker in tomato linked to the *Fusarium oxysporum* resistance gene *I2*. *Theor. Appl. Genet.*, 78, 755-759.

Sari-Gorla, M., Pe, M.D., Mulcahy, D.L., Ottaviano, E. 1992. Genetic dissection of pollen competitive ability in maize. *Heredity*, 69, 423-430.

SAS. 1988. *SAS/STAT User's Guide*, version 6.03. SAS Institute, Cary, North Carolina.

Sauer, S., Lechner, D., Berlin, K., Lehrach, H., Escary, J.L., Fox, N., and Gut, I.G. 2000. A novel procedure for efficient genotyping of single nucleotide polymorphisms. *Nucl. Acids Res.*, 28(5), e13.

Sauer, S., Gelfand, D.H., Boussicault, F., Bauer, K., Reichert, F., Gut, I., 2002. Facile method for automated genotyping of single nucleotide polymorphisms by mass spectrometry. *Nucl. Acids Res.* 30, e22.

Sax, K. 1923. The association of size differences with seed coat pattern and pigmentation in *Phaseolus vulgaris*. *Genetics*, 8, 552-560.

Schmidt, R., West, J., Love, K., Lenehan, Z., Lister, C., Thompson, H., Bouchez, D., and Dean, C. 1995. Physical map and organization of *Arabidopsis thaliana* chromosome 4. *Science*, 270, 480-483.

Schon, C.C., Lee, M., Melchinger, A.E., Guthrie, W.D., and Woodman, W.L. 1993. Mapping and characterization of quantitative trait loci affecting resistance against second-generation European corn borer in maize with the aid of RFLPs. *Heredity*, 70, 648-659.

Schon, C.C., Melchinger, A., Boppenheimer, J., Brunklaus-Jung, E., Herrmann, R.G., and Seitzer, J.F. 1994. RFLP mapping in maize: quantitative trait loci affecting testcross performance of elite European flint lines. *Crop Sci.*, 34, 378-388.

Schondelmaier, J., Steinrucken, G., and Jung, C. 1996. Integration of AFLP markers into a linkage map of sugar beet (*Beta vulgaris* L.). *Plant Breeding*, 115, 231-237.

Scribner, K.T., Arntzen, J.W., and Burke, T. 1994. Comparative analysis of intra- and interpopulation genetic diversity in *Bufo bufo* using allozyme, single-locus microsatellite, minisatellite and multilocus minisatellite data. *Mol. Biol. Evol.*, 11(5), 737-748.

Sekhon, M.S., and Gupta, V.P. 1995. Genetic distance and heterosis in Indian mustard: developmental isozymes as indicators of genetic relationships. *Theor. Appl. Genet.*, 91, 1148-1152.

Senior, M.L., and Heun, M. 1993. Mapping maize microsatellites and polymerase chain reaction confirmation of the targeted repeats using a CT primer. *Genome*, 36, 884-889.

Sharp, P.J., Kreis, M., Shewry, P.R., and Gale, M.D. 1988. Location of β-amylase sequences in wheat and its relatives. *Theor. Appl. Genet.*, 75, 286-290.

Shriver, M.D., Li, J., Chakraborty, R., and Boerwinkle, E. 1993. VNTR allele frequency distributions under the stepwise mutation model: a computer simulation. *Genetics*, 134, 983-993.

Simpson, E.H. 1949. Measurement of diversity. *Nature*, 163, 688.

Slatkin, M. 1991. Inbreedingcoefficients and coalescence times. *Genet. Res. Camb.*, 58, 167-175.

Slatkin, M. 1995. A measure of population subdivision based on microsatellite allele frequencies. *Genetics*, 139, 457-462.

Slocum, M.K., Figdore, S.S., Kennard, W.C., Suzuki, J.Y., and Osborn, T.C. 1990. Linkage arrangement of restriciton fragment length polymorphism loci in *Brassica oleracea. Theor. Appl. Genet.*, 80, 57-64.

Smith, J.S.C., and Smith, O.S. 1989. The use of morphological, biochemical and genetic characteristics to measure distance and to test for minimum distance between inbred lines of maize (*Zea mays* L.). UPOV workshop, Vesailles, October 1989.

Smith, O.S., Smith, J.S.C., Bowen, S.L., Tenborg, R.A., and Wall, S.J. 1990. Similarities among a group of elite maize inbreds as measured by pedigree, F_1 grain yield, grain yield, heterosis and RFLPs. *Theor. Appl. Genet.*, 80, 833-840.

Smith, J.S.C., Smith, O.S., Bowen, S.L., Tenborg, R.A., and Wall, S.J. 1991. The description and assessment of distances between inbred lines of maize. III. A revised scheme for the testing of distinctiveness beween inbred lines utilizing DNA RFLPs. *Maydica*, 36, 213-226.

Smith, J.S.C., Zabeau, M., and Wright, S. 1993. Associations among inbred lines as revealed by RFLPs and by a thermocycling methodology, amplified fragment length polymorphisms (AFLPs). *Maize Genet. Newsl.*, 68, 62-64.

Smouse, P.E., Dyer, R.J., Westfall, R.D., and Sork, V.L. 2001. Two-generation analysis of pollen flow across a landscape. I. Male gamete heterogeneity among females. *Evolution*, 55, 260-271.

Soller, M., and Beckmann, J.S. 1983. Genetic polymorphism in varietal identification and genetic improvement. *Theor. Appl. Genet.*, 67, 25-33.

Soller, M., Brody, T., and Genizi, A. 1976. On the power of experimental designs for the detection of linkage between marker loci and quantitative loci in crosses beween inbred lines. *Theor. Appl. Genet.*, 47, 35-39.

Soltis, D.E., and Soltis, P.S., eds. 1989. *Isozymes in Plant Biology*. Chapman and Hall, 259 pp.

Song, K.M., Suzuki, J.Y., Slocum, M.K., Williams, P.H., and Osborn, T.C. 1991. A linkage map of *Brassica rapa* (syn. *campestris*) based on restriction fragment length polymorphism loci. *Theor. Appl. Genet.*, 82, 296-304.

Sourdille, P., Perretant, M.R., Charmet, G., Leroy, P., Gautier, M.F., Jourier, P., Nelson, J.C., Sorrells, M.E., and Bernard, M. 1996. Linkage between RFLP markers and genes affecting kernel hardness in wheat. *Theor. Appl. Genet.*, 93, 580-586.

Southern, E.M. 1975. Detection of specific sequences among DNA fragments separated by gel electrophoresis. *J. Mol. Biol.*, 98, 503-517.

Sprague, G.F., and Tatum, L.A. 1942. General vs. specific combining ability in single crosses of corn. *J. Am. Soc. Agron.*, 34, 923-932.

Stam, P. 1993. Construction of integrated genetic linkage maps by means of a new computer package: JoinMap. *Plant J.*, 3, 739-744.

Stam, P., and Zeven A.C. 1981. The theoretical proportion of the donor genome in near-isogenic lines of self-fertilizers bred by backcrossing. *Euphytica*, 30, 227-238.

Steffenson, B.J., Hayes, P.M., and Kleinhofs, A. 1996. Genetics of seedling and adult plant resistance to net blotch (*Pyrenophora teres*) and spot blotch (*Cochliobolus sativus*) in barley. *Theor. Appl. Genet.*, 92, 552-558.

Stougaard, J. 2001. Genetics and genomics of root symbiosis. *Curr. Opin. Plant Biol.*, 4, 328-335.

Streiff, R., Ducousso, A., Lexer, C., Steinkellner, H., Gloessl, J., and Kremer, A. 1999. Pollen dispersal inferred from paternity analysis in a mixed oak stand of *Quercus robur* L. and *Q. petraea* (Matt.) Liebl. *Mol. Ecol.*, 8, 831-841.

Stromberg, L.D., Dudley, J.W., and Rufener, G.K. 1994. Comparing conventional early generation selection with molecular marker assisted selection in maize. *Crop Sci.*, 34, 1221-1225.

Stuber, C.W. 1989. Marker-based selection for quantitative traits. *Vortr. Pflanzenzuchtg*, 16, 31-49.

Stuber, C.W. 1995. Mapping and manipulating quantitative traits in maize. *Trends Genet.*, 11, 477-481.

Stuber, C.W., Edwards, M.D., and Wendel, J.F. 1987. Molecular-marker-facilitated investigations of QTL in maize. II. Factors influencing yield and its component traits. *Crop Sci.*, 27, 639-648.

Stuber, C.W., Lincoln, S.E., Wolff, D.W., Helentjaris, T., and Lander, E.S. 1992. Identification of genetic factors contributing to heterosis in a hybrid from two elite maize inbred lines using molecular markers. *Genetics*, 132, 823-839.

Stuber, C.W., and Sisco, P.H. 1993. Marker-facilitated transfer of QTL alleles between elite inbred lines and responses in hybrids. 46th Annual Corn and Sorghum Research Conference, pp. 104-113.

Sturtevant, A.H. 1913. The linear arrangement of six sex-linked factors in *Drosophila*, as shown by their mode of association. *J. Exp. Zool.*, 14, 43-59.

Sun, G.L., Díaz, O., Salomon, B., and von Bothmer, R. 1998. Microsatellite variation and its comparison with allozyme and RAPD variation in *Elymus fibrosus* (Schrenk) Tzvel. (Poaceae). *Hereditas*, 129, 275-282.

Szmidt, A.E., Wang, X.R., and Lu, M.Z. 1996. Empirical assessment of allozyme and RAPD variation in *Pinus sylvestris* (L.) using haploid tissue analysis. *Heredity*, 76, 412-420.

Tajima, F. 1993. Measurement of DNA polymorphism. In *Mechanisms of Molecular Evolution*, Takahata, N., and Clark, A.G., eds., Sinauer Associates, pp. 37-59.

Tajima, F., and Nei, M. 1984. Estimation of evolutionary distance beween nucletotide sequences. *Mol. Biol. Evol.*, 1, 269-285.

Tanhuanpaa, P.K., Vilkki, J.P., and Vilkki, H.J. 1995. Identification of a RAPD marker for palmitic acid concentration in the seed oil of spring turnip rape (*Brassica rapa* ssp. *oleifera*). *Theor. Appl. Genet.*, 91, 477-480.

Tanhuanpaa, P.K., Vilkki, J.P., and Vilkki, H.J. 1996. Mapping of a QTL for oleic acid concentration in spring turnip rape (*Brassica rapa* ssp. *oleifera*). *Theor. Appl. Genet.*, 92, 952-956.

Tanksley, S.D.. 1993. Mapping polygenes. *Annu. Rev. Genet.*, 27, 205-233.

Tanksley, S.D., Bernatzky, R., Lapitan, N.L., and Prince, J.P. 1988. Conservation of gene repertoire but not gene order in pepper and tomato. *Proc. Natl. Acad. Sci. USA*, 85, 6419-6423.

Tanksley, S.D., Ganal, M.W., Prince, J.P., de Vicente, M.C., Bonierbale, M.W., Broun, P., Fulton, T.M., Giovannoni, J.J., Grandillo, S., Martin, G.B., Messeguer, R., Miller, L., Paterson, A.H., Pinedo, O., Roder, M.S., Wing, R.A., Wu, W., and Young, N.D. 1992. High density molecular linkage maps of the tomato and potato genomes. *Genetics*, 132, 1141-1160.

Tanksley, S.D., Ganal, M.W., and Martin, G.B. 1995. Chromosome landing: a paradigm for map-based gene cloning in plants with large genomes. *Trends Genet.*, 11, 63-68.

Tanksley, S.D., Grandillo, S., Fulton, T.M., Zamir, D., Eshed, Y., Petiard, V., Lopez, J., and Beckbunn, T. 1996. Advanced backcross QTL analysis in a cross beween an elite processing line of tomato and its wild relative *L. pimpinellifolium*. *Theor. Appl. Genet.*, 92, 213-224.

Tanksley, S.D., and Hewitt, J. 1988. Use of molecular markers in breeding for soluble solids content in tomato: a re-examination. *Theor. Appl. Genet.*, 75, 811-823.

Tanksley, S.D., Medina-Filho, H., and Rick, C.M. 1982. Use of naturally-occurrin enzyme variation to detect and map genes controlling quantitative traits in an interspecific backcross of tomato. *Heredity*, 49, 11-25.

Tanksley, S.D., and Nelson, J.C. 1996. Advanced backcross QTL analysis: a method for the simultaneous discovery and transfer of valuable QTLs from unadapted germplasm into elite breeding lines. *Theor. Appl. Genet.*, 92, 191-203.

Tanksley, S.D., and Orton, T.J. 1983. Isozymes in plant genetics and breeding. Elsevier, Part A, 516 pp.

Tautz, D. 1989. Hypervariability of simple sequences as a general source for polymorphic DNA markers. *Nucl. Acid Res.*, 17, 6463-6471.

Tersac, M., Blanchard, P., Brunel, D., and Vincourt, P. 1994. Relations between heterosis and enzymatic polymorphism in populations of cultivated sunflowers (*Helianthus annuus* L.). *Theor. Appl. Genet.*, 88, 49-55.

Tessier, N., Bernatchez, L., Presa, P., and Angers, B, 1995. Gene diversity analysis of mitochondrial DNA, microsatellites and allozymes in landlocked Atlantic salmon. *J. Fish Biol.*, 47 (Supplement A), 156-163.

Thoday, J.M. 1961. Location of polygenes. *Nature*, 191, 368-370.

Thomas, B.R., MacDonald, S.E., Hicks, M., Adams, D.L., and Hodgetts, R.B. 1999. Effects of reforestation methods on genetic diversity of lodgepole pine: an assessment using microsatellite and randomly amplified polymorphic DNA. *Theor. Appl. Genet.*, 98, 793-801.

Thomas, M.R., Cain, P., and Scott, N.S. 1994. DNA typing of grapevine: A universal methodology and database for describing cultivars and evaluating genetic relatedness. *Plant Mol. Biol.*, 25, 939-949.

Thomas, W.T.B., Powell, W., Waugh, R., Chalmers, K.J., Barua, U.M., Jack, P., Lea, V., Forster, B.P., Swanston, J.S., Ellis, R.P., Hanson, P.R., and Lance, R.C.M. 1995. Detection of quantitative trait loci for agronomic, yield, grain and disease characters in spring barley (*Hordeum vulgare* L.) *Theor. Appl. Genet.*, 91, 1037-1047.

Thormann, C.E., Romero, J., Mantet, J., and Osborn, T.C. 1996. Mapping loci controlling the concentrations of erucic and linolenic acids in seed oil of *Brassica napus* L. *Theor. Appl. Genet.*, 93, 282-286.

Timmerman-Vaughan, G.M., McCallum, J.A., Frew, T.J., Weeden, N.F., and Russell, A.C. 1996. Linkage mapping of quantitative trait loci controlling seed weight in pea (*Pisum sativum* L.). *Theor. Appl. Genet.*, 93, 431-439.

Torres, A.M., Weeden, N.F., and Martin, A. 1993. Linkage among isozyme, RFLP and RAPD markers in *Vicia faba*. *Theor. Appl. Genet.*, 85, 937-945.

Toth G., Gaspari Z., and Jurka J. 2000. Microsatellites in different eukaryotic genomes: survey and analysis. *Genome Res.*, 10, 967-981.

Touzet, P., Morin, C., Damerval, C., Le Guilloux, M., Zivy, M., and de Vienne, 1995a. Characterizing allelic proteins for genome mapping in maize. *Electrophoresis*, 16, 1289-1294.

Touzet, P., Winkler, R.G., and Helentjaris, T. 1995b. Combined genetic and physiological analysis of a locus contributing to quantitative variation. *Theor. Appl. Genet.*, 91, 200-205.

Tragoonrung, S., Kanizin, V., Hayes, P.M., and Blake, T.K. 1992. Sequence tagged-site-facilitated PCR for barley genome mapping. *Theor. Appl. Genet.*, 84, 1002-1008.

Tulsieram, L.K., Glaubitz, J.C., Kiss, G., and Carlson, J.E. 1992. Single tree genetic linkage analysis in conifers using haploid DNA from megagametophytes. *BioTechnology*, 10, 686-690.

Vallejos, C.E., Sakiyama, N.S., and Chase, C.D. 1992. A molecular marker-based linkage map of *Phaseolus vulgaris* L. *Genetics*, 131, 733-740.

Van den Berg, J.H., Ewing, E.E., Plaisted, R.L., McMurry, S., and Bonierbale, M.W. 1996. QTL analysis of potato tuberization. *Theor. Appl. Genet.*, 93, 307-316.

Van den Berg, J.H., Ewing, E.E., Plaisted, R.L., McMurry, S., and Bonierbale, M.W. 1996. QTL analysis of potato tuber dormancy. *Theor. Appl. Genet.*, 93, 317-324.

Van den Broeck, D., Maes, T., Sauer, M., Zethof, J., Keukeleire, P.D., D'hauw, M., Van Montagu, M., Gerats, T., 1998, *Transposon Display* identifies individual transposable elements in high copy number lines. *Plant J.* 13, 121-129.

Van der Beek, J.G., Verkerk, R., Zabel, P., Lindhout, P. 1992. Mapping strategy for resistance genes in tomato based on RFLPs between cultivars: *Cf9* (resistance to *Cladosporium fulvum*) on chromosome 1. *Theor. Appl. Genet.*, 84, 106-112.

Van Hintum, T.J.L. 1994. Comparison of marker systems and construction of a core collection in a pedigree of European spring barley. *Theor. Appl. Genet.*, 89, 991-997.

Veldboom, L.R., and Lee, M. 1994. Molecular marker facilitated studies of morphological traits in maize. II. Determination of QTLs for grain yield and yield components. *Theor. Appl. Genet.*, 89, 451-458.

Vendramin, G.G., Lelli, L., Rossi, P., and Morgante, M. 1996. A set of primers for the amplifcation of 20 chloroplast microsatellites in Pinaceae. *Mol. Ecol.*, 5, 595-598.

Vicario, F., Vendramin, G.G., Rossi, P., Liò, P., and Giannini, R. 1995. Allozyme, chloroplast DNA and RAPD markers for determining genetic relationships between *Abies alba* and the relic population of *Abies nebrodensis*. *Theor. Appl. Genet.*, 90, 1012-1018.

Vierling, R.A., Faghihi, J., Ferris, V.R., and Ferris, J.M. 1996. Association of RFLP markers with loci conferring broad-based resistance to the soybean cyst nematode (*Heterodera glycines*). *Theor. Appl. Genet.*, 92, 83-86.

Vizir, I.Y., and Korol, A.B. 1990. Sex difference in recombination frequency in *Arabidopsis*. *Heredity*, 65, 379-383.

Vos, P., Hogers, R., Bleeker, M., Reijans, M., van de Lee, T., Hornes, M., Frijters, A., Pot, J., Peleman, J., Kupier, M., and Zabeau, M. 1995. AFLP: a new technique for DNA fingerprinting. *Nucl. Acids Res.*, 23, 4407-4414.

Wang, D.G., Fan, J.B., Siao, C.J., Berno, A, Young, P., Sapolsky, R., Ghandour, G., Perkins, N., Winchester, E., Spencer, J., Kruglyak, L., Stein, L., Hsie, L., Topaloglou, T., Hubbell, E., Robinson, E., Mittmann, M., Morris, M.S., Shen, N., Kilburn, D., Rioux, J., Nusbaum, C., Rozen, S., Hudson, T.J., Lipshutz, R., Chee, M., and Lander, E.S. 1998. Large-scale identification, mapping, and genotyping of single-nucleotide polymorphisms in the human genome. *Science*, 280, 1077-1082.

Wang, G.L., and Paterson, A.H. 1994. Assessment of DNA pooling strategies for mapping of QTLs. *Theor. Appl. Genet.*, 88, 355-361.

Wang, J., and Bernardo, R. 2000. Variance of marker estimates of parental contributions to F_2 and BC1 derived inbreds. *Crop Sci.*, 40, 659-665.

Wang, M.L., Atkinson, M.D., Chinoy, C.N., Devos, K.M., and Gale, M.D. 1992. Comparative RFLP-based genetic maps of barley chromosome 5 1H and rye chromosome 1R. *Theor. Appl. Genet.*, 84, 339-344.

Wang, Y.H., Thomas, C.E., and Dean, R.A. 1997. A genetic map of melon (*Cucumis melo* L.) based on amplified fragment length polymorphism (AFLP) markers. *Theor. Appl. Genet.*, 95, 791-798.

Ward, E.R., and Jen, G.C. 1990. Isolation of single-copy-sequence clones from a yeast artificial chromosome library of randomly-sheared *Arabidopsis thaliana* DNA. *Plant Mol. Biol.*, 14, 561-568.

Ward, J.H. 1963. Hierarchical grouping to optimize an objective function. *J. Am. Stat. Assoc.*, 58, 236-244.

Waugh, R., McLean, K., Flavell, A.J., Pearce, S.R., Kumar, A., Thomas, B.B.T., and Powell, W. 1997. Genetic distribution of Bare-1-like retrotransposable elements in the barley genome revealed by sequence specific amplification polymorphism (S-SAP). *Mol. Gen. Genet.*, 253, 687-694.

Weeden, N.F., Hemmat, M., Lawson, D.M., Loghi, M., Manganaris, A.G., Reish, B.I., Brown, S.K., and Ye, G.N. 1994. Development and application of molecular marker linkage maps in woody fruit crops. *Euphytica*, 77, 71-75.

Weeden, N.F., Muehlbauer, F.J., and Ladizinsky, G. 1992. Extensive conservation of linkage relationships beween pea and lentil genetic maps. *J. Hered.*, 83, 123-129.

Weir, B.S. 1990. *Genetic Data Analysis.* Sinauer Associates, New York, 377 pp.

Weir, B.S., and Basten, J.C. 1990. Sampling strategies for distances. *Biometrics*, 46, 551-582.

Weir, B.S., and Cockerham, C.C. 1984. Estimating F-statistics for the analysis of population structure. *Evolution*, 38, 1358-1370.

Weising, K., and Gardner, R.C. 1999. A set of conserved PCR primers for the analysis of simple sequence repeat polymorphisms in chloroplast genomes of dicotyledonous angiosperms. *Genome*, 42, 9-19.

Weller, J.I. 1987. Mapping and analysis of quantitative trait loci in *Lycopersicon* (tomato) with the aid of genetic markers using approximate maximum likelihood methods. *Heredity*, 59, 413-421.

Welsh, J., and McClelland, M. 1990. Fingerprinting genomes using PCR with arbitrary primers. *Nucl. Acids Res.*, 19, 861-866.

Whitkus, R., Doebley, J., and Lee, M. 1992. Comparative genome mapping of sorghum and maize. *Genetics*, 132, 1119-1130.

Whitlock, M.C., and McCauley, D.E. 1990. Some population genetic consequences of colony formation and extinction: genetic correlations within founding groups. *Evolution*, 44, 1717-1724.

Williams, J.G.K., Kubelik, A.R., Livak, K.J., Rafalski, J.A., and Tingey, S.V. 1990. DNA polymorphism amplified by arbitrary primers are useful as genetic markers. *Nucl. Acids Res.*, 18, 6531-6535.

Wolff, K., Rogstad, S.H., and Schaal, B.A. 1994. Population and species variation of minisatellite DNA in *Plantago*. *Theor. Appl. Genet.*, 87, 733-740.

Wright, S. 1951. The genetical structure of populations. *Ann. Eugen.*, 15, 323-354.

Wright, S. 1968. *Evolution and the Genetics of Populations*, vol. 1. *Genetics and Biometric Foundations*. The University of Chicago Press.

Wright, S. 1969. *Evolution and the Genetics of Populations*, vol. 2. *The Theory of Gene Frequencies*. The University of Chicago Press.

Wu, J., Krutovskii, K.V., and Strauss, S.H. 1999. Nuclear DNA diversity, population differentiation, and phylogenetic relationships in the California closed-cone pines based on RAPD and allozyme markers. *Genome*, 42, 893-908.

Wu, K.S., and Tanksley, S.D. 1993. Abundance polymorphism and genetic mapping of microsatellites in rice. *Mol. Gen. Genet.*, 241, 225-235.

Xiao, J., Li, J., Yuan, L., McCouch, S.R., and Tanksley, S.D. 1996a. Genetic diversity and its relationship to hybrid performance and heterosis in rice as revealed by PCR-based markers. *Theor. Appl. Genet.*, 92, 647-643.

Xiao, J., Li, J., Yuan, L., and Tanksley, S.D. 1996b. Identification of QTLs affecting traits of agronomic importance in a recombinant inbred population derived from a subspecific rice cross. *Theor. Appl. Genet.*, 92, 230-244.

Yamamoto, K., and Sasaki, T. 1997. Large-scale EST sequencing in rice. *Plant Mol. Biol.*, 35, 135-144.

Yan, G., Romero-Severson, J., Walton, M., Chadee, D.D., and Severson, D.W. 1999. Population genetics of the yellow fever mosquito in Trinidad: comparisons of amplified fragment length polymorphism (AFLP) and restriction fragment length polymorphism (RFLP) markers. *Mol. Ecol.*, 8, 951-963.

Yang, G.P., Saghai Maroof, M.A., Xu, C.G., Zhang, Q, and Biyashev, R.M. 1994. Comparative analysis of microsatellite DNA polymorphism in landraces and cultivars of rice. *Mol. Gen. Genet.*, 245, 187-194.

Yano, M., Katayose, Y., Ashikari, M., Yamanouchi, U., Monna, L., Fuse, T., Baba, T., Yamamoto, K., Umehara, Y., Nagamura, Y., and Sasaki, T. 2000. *Hd1*, a major photoperiod sensitivity quantitative trait locus in rice, is closely related to the *Arabidopsis* flowering time gene *CONSTANS*. *Plant Cell*, 12, 2473-2484.

Ye, S., Dhillon, S., Ke, X., Collins, A.R., and Day, I.N.M. 2001. An efficient procedure for genotyping single nucleotide polymorphisms. *Nucl. Acids Res.*, 29(17), 1-8.

Young, N.D. 2000. A cautiously optimistic vision for marker-assisted breeding. *Mol. Breeding*, 505-510.

Young, N.D., Danesh, D, Menancio-Hautea, D., and Kumar, L. 1993. Mapping oligogenic resistance to powdery mildew in mungbean with RFLPs. *Theor. Appl. Genet.*, 87, 243-249.

Young, N.D., and Tanksley, S.D. 1989. RFLP analysis of the size of the chromosomal segments retained around the *Tm-2* locus of tomtato during back-cross breeding. *Theor. Appl. Genet.*, 77, 353-359.

Young, N.D., Zamir, D., Ganal, M.W., and Tanksley, S.D. 1988. Use of isogenic lines and simultaneous probing to identify DNA markers tightly linked to the *Tm-2a* gene in tomato. *Genetics*, 120, 579-585.

Yu, Z.H., MacKill, D.J., Bonman, J.M., and Tanksley, S.D. 1991. Tagginggenes for blast resistance in rice via linkage to RFLP markers. *Theor. Appl. Genet.*, 81, 471-176.

Zamir, D., Ekstein-Michelson, I., Zakaiy, Y., Navot, N., Zeidan, M., Sarfatti, M., Eshed, Y., Harel, E., Pleban, T., Van Oss, H., Kedar, N., Rabinowitch, H.D., and Czonek, M. 1994. Mapping and introgression of a tomato yellow leaf curl virus tolerance. *TY-1. Theor. Appl. Genet.*, 88, 141-146.

Zanetto, A., Roussel, G., and Kremer, A. 1994. Geographic variation of inter-specific differentiation between *Quercus robur* L., *Quercus petraea* (Matt.) Liebl. *Forest Genet.*, 1, 111-123.

Zehr, B.E., Dudley, J.W., Chojeki, J., Maroof, M.A.S., and Mowers, R.P. 1992. Use of RFLP markers to search for alleles in a maize population for improvement of an elite hybrid. *Theor. Appl. Genet.*, 83, 903-911.

Zeng, Z.B. 1993. Theoretical basis for separation of multiple linked gene effects in mapping quantitative trait loci. *Proc. Natl. Acad. Sci. USA*, 90, 10972-10976.

Zeng, Z.B. 1994. Precision mapping of quantitative trait loci. *Genetics*, 136, 1457-1468.

Zhang, Q., Gao, Y.J., Li, J.X., Saghai Maroof, M.A., and Yang, S.H. 1995. Molecular divergence and hybrid performance in rice. *Mol. Breeding*, 1, 133-142.

Zhang, Q., Gao, Y.J., Yang, S.H., Ragab, R.A., Saghai Maroof, M.A., and Li, Z.B. 1994. A diallel analysis of heterosis in elite hybrid rice based on RFLPs and microsatellites. *Theor. Appl. Genet.*, 89, 185-192.

Zhang, Q, Saghai Maroof, M.A., and Kleinhofs, A. 1993. Comparative diversity analysis of RFLPs and isozymes within and among populations of *Hordeum vulgare* ssp. *spontaneum*. *Genetics*, 134, 909-916.

Zhang, W., and Smith, C. 1992. Computer simulation of marker-assisted selection utilizing linkage disequilibrium. *Theor. Appl. Genet.*, 83, 813-820.

Zhu, Y., Strassmann, J.E., Queller, D.C., 2000. Insertions, substitutions, and the origin of microsatellites. *Genet. Res.* 76, 227-236.

Zietkiewicz, E., Rafalski, A., and Labuda, D. 1994. Genome fingerprinting by simple-sequence repeat (SSR)-anchored polymerase chain reaction amplification. *Genomics*, 20, 176-183.

Zivy, M., Devaux, P., Blaisonneau, J., Jean, R., and Thiellement, H. 1992. Segregation and linkage studies in microspore-derived double haploid lines of *Hordeum vulgare, L. Theor. Appl. Genet.*, 83, 919-924.

Index

For Product Safety Concerns and Information please contact our EU
representative GPSR@taylorandfrancis.com Taylor & Francis Verlag GmbH,
Kaufingerstraße 24, 80331 München, Germany

Printed and bound by CPI Group (UK) Ltd, Croydon, CR0 4YY
02/05/2025
01859636-0001